Above: A Paddington-Bristol express with No 6008 *King James I*, 1929. *Ian Allan Library*

DECADES OF STEAM 1920-1969

Michael Harris

Contents

Front cover, top left: LNER 'A4' Pacific No 2509 *Silver Link* ready to leave Kings Cross with the down 'Silver Jubilee' in April 1937. *C. S. Perrier collection/Colour-Rail*

Front cover, top right: BR Standard Class 4 No 76089 seen at New Mills South Junction when brand-new. *W. Oliver/Colour-Rail*

Front cover, bottom right: Ex-GWR 'King' No 6006 *King George I* with the down 'Bristolian' near Thingley Junction in 1954. *P. M. Alexander/Colour-Rail*

Front cover, bottom left: Southern Railway 'Merchant Navy' No 21C12 *United States Lines* at Queens Road on the down 'Bournemouth Belle' in April 1947. *H. N. James/Colour-Rail*

Back cover, top: Ex-LMS 'Princess Royal' Pacific No 46203 *Princess Margaret Rose* approaches Kensal Green Tunnel with the Holyhead-Euston 'Emerald Isle Express' on 10 September 1960. *K. L. Cook/Rail Archive Stephenson*

Back cover, bottom: Ex-LNER 'J39' No 64733 near Hawick with an up freight over the Waverley route in September 1959. *Colour-Rail*

Title page: The northbound 'Devonian' express, north of Bristol, May 1955. *George Heiron*

Below: 'Star' No 4021 *King Edward* prepared to work a royal train. *Ian Allan Library*

First published 1999

ISBN 0 7110 2683 1

© Ian Allan Publishing Ltd 1999

Published by Ian Allan Publishing

an imprint of Ian Allan Publishing Ltd,
Terminal House, Shepperton, Surrey TW17 8AS.
Printed by Ian Allan Printing Ltd,
Riverdene Business Park, Hersham, Surrey KT12 4RG.

Code: 9910/B2

Introduction

Decades of Steam - yes, for several decades steam railways and the use of steam locomotives were taken for granted. Even when diesel and electric locomotives were entering service in large numbers, there seemed to be some corner on British Railways (BR) where the steam railway still survived. While you might arrive at Newcastle Central behind a high-powered 'Deltic' main line diesel, not far away, at South Blyth perhaps, there were still ex-North Eastern Railway (NER) 0-6-0s undertaking the same duties that they had carried out for at least two generations of railwaymen. The railway environment in so many places remained that of the steam age.

The history of railways is not however the history of steam traction. The machinations of the companies before and after the Grouping were on a much broader, grander scale. They were involved in so many different businesses - shipping, road transport, hotels, docks and harbours, canals - just to name the main headings. The ending of steam traction in 1968

did not in fact change the railway environment and the way in which railways functioned as much as the various waves of change - technical and managerial - that were to transform our railways in the period from 1975.

This book is an accompaniment to the publisher's SBS Video series, *Decades of Steam.* It reflects some of the themes covered in the five videos in the series - covering the steam railway of the 1920s, 1930s, 1940s, 1950s and 1960s. Like the videos, the book is selective in its treatment. I have attempted to look at some of the aspects of operating the steam age railway and, in the case of the 1950s and 1960s, how BR made the transition to a system that no longer relied on steam traction.

There is an emphasis on the steam locomotives themselves and when, why and how changes were made to the fleets of engines that so fascinated many

Above:
Wheeling an ex-NER 'A8' 4-6-2T during World War 2.
Ian Allan Library

people. Their fascination with motive power meant that enthusiasts tended to regard railway chief mechanical engineers as on a par with general managers or company chairmen. If you work your way through the surviving archives of our railways or even glance at the official railway industry handbooks, the Chief Mechanical Engineer (CME) is by no means at the top of the list of railway chief officers. Of those officials attending a meeting of the London & North Eastern's (LNER) Joint Locomotive & Traffic Committee in 1937, Sir Nigel Gresley appears ninth in the list. With Nationalisation, the senior mechanical engineers concerned with policy of motive power seemed to have enjoyed a high profile, and perhaps to have received more publicity.

In *Decades of Steam 1920-1969* I have made use of as many primary sources as possible, if for no other reason than to try to offer some fresh perspectives on an otherwise familiar story. The information is there in our national archives and there are a few surprises still to be found. I have tried to include some in the book, relating variously to the enterprise of the LNER, new angles on the Big Four in the 1930s, to twists and turns in policies connected with the Modernisation Plan, and as highlights of different aspects of the rundown of the steam age railway during the 1960s. Overall, any mistakes are attributable to me alone, and have no bearing on the *Decades of Steam* videos.

Michael Harris
Ottershaw, Surrey
May 1999

Below:
Steam, Gowhole Sidings, 1967. *R. I. Vallance*

1. Dawning of a New Age:
The Challenge to the Steam Railway's Supremacy

Many will say that British railways were at their peak in the years before the outbreak of World War 1. Yet, although on paper their competence, earning capacity and generally good condition were at an all-time high, beneath the surface matters were not so golden. Industrial unrest had been a feature of the first years of the 1910s. Early in 1912 the coal strike caused the railways an appreciable loss of traffic and increased their working costs. This dispute had followed the damaging railway strike of August 1911. Then the wet summer of 1912 cut the number of holidaymakers travelling by train. Recent attempts by some of the railway companies to combine into groups had been resisted by Government and the growth of competitive forms of transport had begun to erode passenger traffic, notably in cities.

With the outbreak of World War 1, the British railways passed into Government control, administered by the Railway Executive Committee which consisted of a number of the general managers. The Committee came under the direction of the Board of Trade which had responsibility in those days for the regulation of railways. So, although the railways' own attempts before 1914 to group themselves into larger units had been resisted by Government, in wartime they were effectively nationalised. They were hard-worked, too. Almost 50% of railway staff of military age joined the armed forces. Locomotives, coaching stock and wagons were pressed into war use, railway workshops were engaged on producing

Above:
The magnificence of British pre-Grouping railways: built in 1909, London Tilbury & Southend Railway 4-4-2T No 80 *Thundersley* was specially decorated for King George V's coronation in June 1911, with garlands and the busts of King and Queen carried above the buffer beam. *LPC/Ian Allan Library*

Below:
Between 1901 and 1908, the London, Brighton & South Coast Railway rebuilt its half of London Victoria station. Here is a view of the platform ends with, on the left: an 'E5' 0-6-2T and right, an 'H1' Atlantic No 40. By the time of this photograph Platforms 1-5 (seen here) had been electrified at 6,600V ac for use by electric trains. *LPC/Ian Allan Library*

military supplies and munitions, and the system as a whole was hard-pressed, with trains delivering personnel and equipment to ports for shipping to the Western Front and remarkable movements of coal transported to fire the boilers of the warships of the Grand Fleet.

As the war to end all wars proceeded, Government pressed the railways to reduce their passenger traffic, and accordingly during 1917 fares were increased and train services reduced. By the time the Armistice had been signed, the British railway system was badly run down, and passenger services, particularly on principal routes, were a pale shadow of their 1914 best, in terms of frequency, speed and attractiveness.

The efficiency and capability of British railways undoubtedly contributed to the war effort and railways were crucial in serving the Allied forces on the Western Front, yet already it was clear that motorised road transport was indispensable for many tasks. In support of the war effort, and helped by Government subsidy, the production of road vehicles had increased dramatically. Many were no doubt regarded as expendable, but enough ex-military lorries remained at the end of hostilities to make it easy for small operators to set up road collection and delivery services. This gave the road freight business a flying start after 1919.

Any opportunities for a flying start were missing from British railways after World War 1. They remained under Government control while arguments flew as to whether they should be nationalised - as the railway trades unions strongly advocated - and/or thoroughly modernised. Government dithered. Such plans were in any case overtaken by other events. In February 1919,

Above:
Elegant and efficient: Churchward's engines and coaches for the Great Western Railway were streets ahead of other pre-Grouping railways. 'Saint' 4-6-0 No 2908 *Lady of Quality* heads a rake of 57ft 'Toplight' corridor stock forming a Paddington-Birmingham express. *LPC/Ian Allan Library*

Right:
Before 1914, railway staff were certainly loyal but increasingly aware of their companies' resistance to trade unionism. This 1907 picture depicts Swindon running shed's improvement committee which would have been run by the enginemen in their own time, both to assist firemen to be passed out as drivers, and to develop their members' general knowledge of locomotives and engineering. The engine in the background is 'Saint' No 2903 *Lady of Lyons. Ian Allan Library*

Enginemen and Firemen's Improvement Class, Swindon, 1907.

Above:
It took the pre-Grouping railways some time to introduce through workings by the engines of one company on to the lines of another. From 1908, the Highland Railway and Great North of Scotland Railway co-operated with the joint working of Inverness-Aberdeen trains, such as the 9.8am from Inverness arriving at Aberdeen Joint station on 4 August 1911 behind Highland 4-4-0 No 2 *Ben Alder* on a rake of GNSR stock. *K. A. C. R.Nunn/ Locomotive Club of Great Britain*

Top left:
Attempts made by the pre-Grouping railways to merge and so form larger groups were mostly frustrated. In August 1912, however, the Midland Railway received Government approval to take over the London, Tilbury & Southend. In Midland livery and numbering, ex-LT&S 4-4-2T No 2178 (LT&S No 81 *Aveley*) is seen leaving Upminster on a Southend train formed of LT&S bogie and six-wheeled stock. *LPC/Ian Allan Library*

Left:
The paint shop of the London, Brighton & South Coast Railway's Brighton Locomotive Works. The engine nearest the photographer is 1900-built 'B4' 4-4-0 No 54 *Empress* which was notable for having hauled Queen Victoria's funeral train. *Madgwick Collection/Ian Allan Library*

Left:
Just before World War 1, the South Eastern & Chatham Railway invested heavily in improving facilities at Dover. 'E' class 4-4-0 No 504, built at Ashford, 1906, is seen leaving the Admiralty Pier extension in 1913.
Ian Allan Library

Left:
Great Northern Railway 2-4-0 No 895 was one of a mixed bag of engines loaned to the South Eastern & Chatham Railway to overcome a locomotive shortage at about the time of World War 1. No 895 was on loan in 1914/15 and during the time of its loan its tender was relettered SE&CR.
Ian Allan Library

Below:
The road approach to Southampton West station showing the transition from horse-drawn to motor vehicles after World War 1. The Leyland lorry is a Royal Air Force vehicle - many such were sold to civilians starting in the road haulage business. *Ian Allan Library*

an eight-hour working day came into effect for railway workers but industrial relations between the trades unions and Government steadily deteriorated, culminating in the nine-day railway strike of September 1919. Road freight and passenger transport helped to undermine the strike. Prices and wage rates rocketed in the early postwar period, until mid-1920.

Although the prospect of nationalisation faded, in influential quarters the received wisdom was that unfettered competition between transport modes should be absent from the postwar world. The railways had been under unified control during wartime and surely, the argument ran, a measure of control should be retained.

The Government set up the Ministry of Transport in 1919, one of its objectives being to control the power of the railways by regulating the fares and rates charged. The new Ministry was much less concerned with the effects of the growth of road transport, which was to weaken the position of the railways and any thought of monopoly. By 1924, the number of goods vehicles on the roads was double that of four years earlier. Stage bus and coach services increased rapidly through the 1920s. Road operators of all kinds enjoyed much greater freedom in handling their traffic than did the railways. It took until 1930 before Government took regulatory action. .

With the collapse of the short-lived postwar economic boom, the railways were well advanced towards their Grouping under the Railways Act of

1921 which received Royal Assent on 19 August 1921,
four days after the railways left wartime Government
control.

Out of 120 main constituents and subsidiary
companies, instead of a single nationalised concern
there were now four large grouped railway
companies - the Great Western (GWR); the Southern
(SR); the London, Midland & Scottish (LMS) and, the
London & North Eastern (LNER). The new structure
came into force on New Year's Day, 1923. At one
stage, there was a proposal for the five Scottish
companies to be grouped into two but in the finalised
Grouping arrangements they were distributed
between the LMS and LNER.

During World War 1, Government pressure on the
railways to cut the total passenger train mileage
operated saw a number of services withdrawn and
stations closed. Many were probably uneconomic
anyway and no effort was made to reinstate them as
the railways recovered from the effects of the war.
One reason was that traffic did not pick up in the face
of unemployment which hit the railways hard,
particularly in their industrial heartlands. Although in
1921 many summertime express trains were either
reinstated for the first time since 1914, or else were
entirely new, the country experienced a severe
industrial depression during 1921 and 1922 and the
railways were gloomy about their prospects.

Throughout the 19th century British railways had
relied on basic and generally unsophisticated designs of
locomotives. This state of affairs changed with the new
century which saw the introduction of large 4-4-0s and
Atlantics to work express trains that rapidly became

Above:
The Midland Railway went in for a 'small engine' policy and worked many passenger and goods trains with pairs of engines. This scene at Kegworth, north of Leicester, dated 1 April 1914, shows a lengthy up coal train worked by Johnson 0-6-0s Nos 3552 and 2904. *Ian Allan Library*

Below:
The north end of Crewe station, c1898 with a selection of modestly powered LNWR engines, including 'Large Jumbo' 2-4-0 No 890 *Sir Hardman Earle* of 1895. It was withdrawn in 1928 as LMS No 5051. Note the horse being used for shunting to the right of the picture. *Ian Allan Library*

heavier with the arrival of more modern bogie stock and the inclusion of catering vehicles or sleeping cars in trains which in addition were running to faster schedules. It took longer for larger goods locomotives, such as 0-8-0s and 2-8-0s, to become widely used.

Late in World War 1 a number of larger locomotive designs were introduced. Some were 2-8-0s, some 2-6-0s, and some were 4-6-0s, the latter as much for mixed traffic haulage as express passenger duties. Not all were successful. With the relatively slow timings of the immediately postwar express trains, some of the mixed traffic engines had no trouble keeping time, despite loads in the 450-600-ton range, particularly during the coal strike of 1921. Other types, such as the Caledonian Pickersgill 4-6-0s were saved embarrassment: they were simply sluggish in running and, judged by the standards of 1914, would have fallen short. With the extended timings of the early 1920s their shortcomings were largely veiled.

Other 4-6-0s, such as those built by the London & South Western Railway (LSWR), North Eastern and Great Central railways (GCR), and the Great Northern Railway's (GNR) 2-6-0s were intended to haul the growing number of fast freight and perishable traffic trains with partial or full through vacuum braking. Some of these types had first made their appearance before World War 1 but were only multiplied later.

Just before and immediately after World War 1 several railways took a close look at the processing of freight traffic. The answer seemed to lie in large marshalling yards which in themselves were nothing new; indeed, some had been built in the 19th century. The new wave were however on a much larger scale.

Above:
An example of the larger engines and rolling stock introduced from the start of the 20th century. This is Great Central '8' (or 'Fish') class 4-6-0 No 1072, a Robinson engine built in 1902. It is standing at the head of some splendid GCR bogie fish wagons of 15-ton capacity, put into service from 1902 for Grimsby fish traffic.

Opposite top:
One of the larger mixed traffic engines dating from the post-World War 1 years: Great Northern Railway 2-6-0 No 1000 of 1920, later LNER 'K3' No 4000, heads a heavy down express from King's Cross through New Southgate.
Ian Allan Library

Opposite:
In July 1920, the Great Eastern Railway brought in what at the time was termed 'The Last Word in Steam-Operated Suburban Train Services' to counter calls for electrification. The frequency and capacity of the train service were boosted as a result of careful planning, but with minimal expenditure. This scene is of the platform ends of the west side of Liverpool Street station.
Ian Allan Library

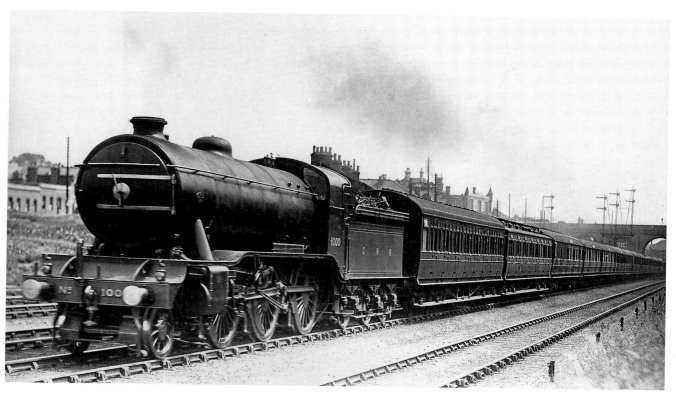

Left:
In contrast with the GER's suburban train services, those operated by the LNWR to destinations on Great Northern Railway territory were essentially Victorian. They were worked by aged North London outside-cylindered 4-4-0Ts and close-coupled rakes of four-wheeled NLR coaches, as seen here near Wood Green, heading towards the City. *H. Gordon Tidey/ Ian Allan Library*

Right:
Before and during World War 1, suburban electrification schemes were completed, such as that involving the London & North Western Railway's London area lines. This is one of the company's comfortable and spacious Oerlikon electric trains dating from 1915-24. *Ian Allan Library*

Below:
Typical of the larger railway goods depots was the Caledonian Railway's Buchanan Street, Glasgow Goods Station, shown with numerous horse-drawn vehicles.
Ian Allan Library

New engines were matched by the limited re-equipment of wagon fleets and coaching stock in the 1918-23 period. The British railway companies were hampered in their freight operations by the existence of private owner wagons. In 1914, the combined fleets were distributed among 4,000 owners and it was clear that the railways did not have the resources to buy them up. The wagons were for the most part of outdated design and often badly maintained so that they impeded efficient freight operations.

While the railways were encountering competition from road transport, they were also becoming large operators in their own right. The railways later moved to acquire financial interests in bus companies, in addition to some companies' bus operations that dated from before World War 1. Railway road collection and delivery services were set up in the 1920s - the powers for which and some obligations to collect and deliver rail-borne traffic having been made clear in the 1921 Railways Act.

Railways were also investing in other properties, notably hotels. Such investment in facilities for the better-off was reflected in a smaller way by the resurgence in Pullman car operations after World War 1. In December 1919, the Great Eastern Railway (GER) and the Pullman Car Co signed a 20-year agreement for the operation of cars on Continental boat trains and other expresses. Although traffic potential was limited in the Eastern Counties, in the later 1920s and 1930s the operating agreement formed the basis for new Pullman services to the West Riding and Scotland. The Caledonian Railway and Pullman Car Co had also co-operated to introduce Pullman cars on to a wide range of internal Scottish services.

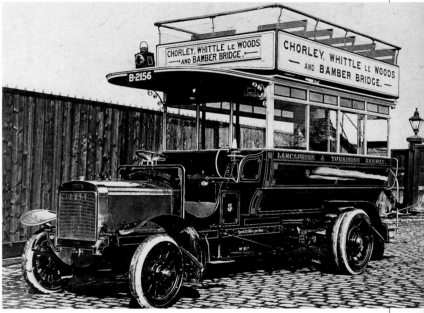

Right:
A Commer 34-seater bus acquired by the Lancashire & Yorkshire Railway in 1908 to operate its Chorley-Bamber Bridge route. The service ceased as a result of poor road conditions and lack of traffic, but this bus survived as a lorry until 1920. *Ian Allan Library*

Below:
The railways made some use of steam road vehicles, such as this Yorkshire solid-tyred steam wagon working from the Great Western Railway's Paddington Goods depot, and seen hauling a converted horse dray. *Ian Allan Library*

Overleaf:
The London & North Western Railway invested heavily in its shipping routes based on Holyhead. Here is a view of the harbour from the station ticket office showing the *Scotia*, delivered in 1921. *Ian Allan Library*

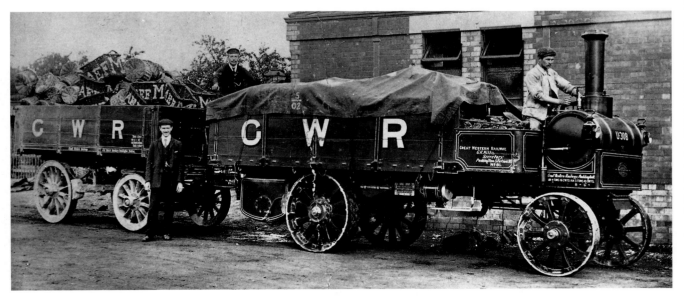

2. **The 1920s:**

From Grouping Towards Depression - from 2-4-0Ts to 'Kings'

Road transport multiplied rapidly through the 1920s such that by 1930 there were three times the number of goods vehicles that had been on the roads in 1920; from less than 200,000 cars in 1920 there were over a million in 1930. The number of buses had increased appreciably: in the four years following the General Strike of 1926 the total had risen by a quarter. On journeys up to 10 miles the railways suffered especially. A survey of LMS rail passenger traffic indicated that on journeys of this distance the loss in receipts was over 25%. Coach services also ate into rail traffic, particularly in the Home Counties, in East Anglia and the Midlands. From 1928 the Big Four companies were given powers to operate their own bus services and although services and bus fleets were

built up over the next few years, their policies then changed. Gradually the railways disposed of their bus fleets - only the Southern had never operated buses on its own account. The Big Four's tactics changed instead to taking financial stakes in the larger bus companies.

On much of British railways aged equipment pottered around. By the mid-1920s there was scope for rationalisation as thinly-trafficked lines lost much of what they had to road. The Great Western Railway looked critically at its branch lines in 1926. The terms of reference of a report prepared by senior management were: to see whether branch lines were

Above:
Euston station and Hardwick's renowned *propylaeum* (popularly known as the Euston Arch). A scene from early LMS days. *Ian Allan Library*

Right:
A railway bus - a Thornycroft 'BC' - put into service by the LNER late in 1928 (as a result of the Railways [Road Transport] Act, 1928) on routes extending from Durham. It was painted in a cream and green livery. *Ian Allan Library*

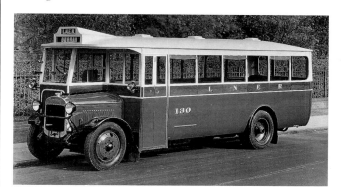

being worked most economically and if the present system was capable of improvement; if the use of steam locomotives could be abolished on some lines and substituted by a steam rail-motor and trailer, and indeed whether on some lines 'the rails be taken up and the line used as a motor road by the company'. To work passenger trains on branches it cost 1s 6¾d using locomotive and coaches, but 1s 4d with a rail-motor. The report looked at receipts and paybills at each station on the Company's branches, the level of traffic and the method of working. Recommendations were made for each branch, such as whether signalboxes might be closed and the scope for making savings in staff positions.

Even by the mid-1920s some railways outside of the Big Four were already an anachronism, none more so than the 2ft 6in gauge Southwold Railway which ran nearly nine miles from Halesworth on the East Suffolk line to Southwold. It had missed being incorporated into the LNER at Grouping and after World War 1 entered its decline. There was no money for improvement of a railway that had been operating largely unchanged for half a century but after the war it carried as many as 85,000 passengers annually and 15,000 tons of freight; more, in fact, than it had

conveyed before 1914. But passenger traffic dropped alarmingly during 1926, in part due to the General Strike which naturally depressed economic activity.

In 1928 the Southwold Railway carried 25,000 less passengers because competing bus service frequencies had been increased. Despite reducing its fares and cutting wages, the Southwold Railway simply could not compete. Early in 1929, the Railway's directors fruitlessly approached the LNER for financial assistance. There was no other option than to close the line completely and the last passenger train was the 5.23pm departure from Southwold on the bitterly cold 11 April 1929. No one had the authority to wind up the company and so the Southwold Railway's possessions were left to rot until in 1941 scrap teams moved in for wartime salvage.

Not all those railways independent of the Big Four were Victorian relics such as the Southwold Railway. Take the Metropolitan Railway, for instance. This London suburban railway with pretences to being a main line operation had introduced its first multiple-unit electric service in 1905, between its Baker Street, London station and Uxbridge. From 1906, the Met's main line trains to Chesham, Aylesbury and the remote outpost at Verney Junction were hauled by

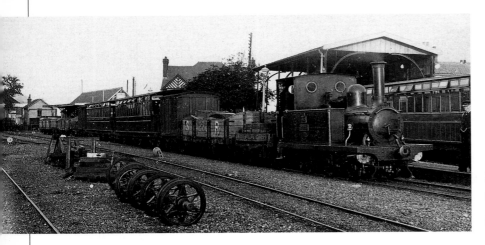

Left:
The 3ft gauge Southwold Railway, at Southwold station, c1912, with 2-4-0T No 3 *Blyth* (built by Sharp, Stewart, 1879) at the head of a mixed train for the junction with the main line at Halesworth. *LPC/Ian Allan Library*

Left:
Main line Metropolitan Railway: one of the 1904 electric locomotives, No 3, with a train that includes one of the two Pullman cars used on the railway from 1910.
Ian Allan Library

Above:
The Great Western Railway's Paddington Goods Depot, coping with a backlog of traffic following the General Strike of 1926. Local deliveries are principally by means of horse-drawn vehicles.
Science & Society Picture Library GWR/B5774

electric locomotives between Baker Street and Wembley Park; electric haulage was later extended to and from Harrow. Baker Street station was extensively reconstructed from 1911. The line between Finchley Road and Wembley was altered to quadruple track by 1915 and, in a joint venture with the LNER, a new branch line was opened from Moor Park to Watford in 1925. This was electrically worked from the start, as was the Wembley Park-Stanmore branch which was completed in December 1932. But the Metropolitan's independence was nearing its end, for an Act of Parliament of April 1933 established the London Passenger Transport Board (LPTB). The Metropolitan had been opposed to the plan for a unified transport undertaking for the capital but from July 1933 the Company, together with the District Railway and associated tube railways, the municipal and company-operated tramways and the principal bus companies, was incorporated into the new undertaking.

The LPTB made it clear that it did not share the Metropolitan's main line aspirations and steadily cut back what were seen as peripheral activities. Thirty-six steam locomotives had been taken over by London Transport; from late 1937, half this fleet passed to the LNER which now operated Metropolitan Line passenger trains north of Rickmansworth and almost all commercial goods operations on the system. Met electric locomotives continued to work south of Rickmansworth, on trains to Baker Street and

the City, and (until 1939) between Paddington and the City at the head of GWR trains from that railway's Thames Valley suburban stations.

The Big Four railways offered extensive collection and delivery services between customers and railway depots. The majority of these were operated with horses. The minutes of the pre-Grouping and Big Four railways make frequent mention of the acquisitions and disposals of horses and the facilities needed to sustain them. For much of the Victorian period the railways had made use of cartage agencies but gradually the railways had taken over the task. At the peak of horse-drawn railway cartage there were 15,000 horses at work, their working life being nine years. With the arrival of motorised road transport the railways first used lorries and vans chiefly in rural areas and for particularly arduous duties in towns. That said, there had been some use of steam lorries from the early years of the century. It was only the development of the articulated mechanical horse from 1930 that allowed horse haulage to be reduced although as late as 1947 the Big Four employed nearly 9,000 horses on cartage work and for shunting wagons.

The GWR in the 1920s

Alone of the Big Four companies, the GWR had retained its old name and identity of pre-Grouping days. As from New Year's Day, 1922, the GWR had gathered-in the Barry, Cambrian, Cardiff, Rhymney, Taff Vale and Alexandra (Newport and South Wales) companies. Between 1922 and July 1923 it also took under its wing a number of other subsidiary companies including the Brecon & Merthyr, Port Talbot, and Midland & South Western Junction railways.

The GWR faced a major challenge in absorbing diverse fleets of locomotives, coaching stock and wagons. Generally, it got to grips quickly with its inheritance although officially it was only in the late summer of 1922 that the GWR added any of the locomotives to stock. It had acquired 720 from the six main Welsh railways and a further 205 from the smaller undertakings. These absorbed engines were renumbered, classified, weeded out when they were clearly of no further use, or placed on a sales list if they were likely to be purchased by another operator, while the balance of the stock was repaired, reboilered and generally brought up to GWR standard. The general tendency though was one of replacement.

Not that all the engines in the existing GWR fleet were necessarily of modern design. Of course, the Churchward era had seen a major restocking of the railway with standardised locomotives, usually of great capability. During the first 10 years or so of his tenure of office as the Locomotive Carriage and Wagon Superintendent, Churchward worked hard to establish designs that were suitable for the principal traffic demands and then refine them for quantity production. In the ensuing 10 years ending in 1922, the standard types built included 'Stars', 'Saints', 2-8-0s, 2-8-0Ts, large and small 2-6-2Ts, the excellent '43xx' 2-6-0s, and the less successful 'County' 4-4-0s and 'County' tanks.

The older engines included the 'Duke' 4-4-0s from Dean's period of office. By now a quarter of a century old, from 1900 'Dukes' had been superseded on their original duties on the main line through Devon and Cornwall and had passed to secondary duties. With Grouping, the 'Dukes' enjoyed a new lease of life for they were drafted to the Cambrian lines to displace the resident 4-4-0s on through passenger trains. The 'Dukes' were also allocated to work the newly

Below:
The Big Four railways were a major force in the operation of Britain's ports and harbours. This is a view of the GWR's Cardiff Bute Docks dated 1927, with coal hoists clearly visible. There were 140 miles of sidings in Bute Docks. *Ian Allan Library*

acquired Midland & South Western Junction line, and replaced GWR 2-4-0s on country routes such as that between Didcot and Winchester.

Similarly, the Dean Goods, an 0-6-0 design that had been multiplied to 260 between 1883 and 1899, had over 20 or so years of useful service ahead of them. During World War 1 nearly a quarter of the total had seen service in France and elsewhere and some never returned home. After that war, the Dean Goods were gradually replaced by new engines like the '43xx' 2-6-0s and a number were concentrated in Central Wales where they handled passenger and goods services, often covering long distances on through workings.

While these relatively old but useful and easy to maintain engines continued at work, the GWR got to grips with clearing out other elderly and less numerous classes. Engines such as the '3521' class 2-4-0s were displaced by new 2-6-2Ts during the late 1920s, while the last survivors of the '3201' or 'Stella' class, and the '3232' 2-4-0s disappeared by the early 1930s. Some of the last survivors were withdrawn from the Chester Division where the final examples of the '3206' or

'Barnum' class lingered until the mid-1930s. These three types of 2-4-0 tender engines had been put into service in the last couple of decades of the 19th century and, although a number had been superheated not long before Grouping, when withdrawn they were well past their 'book' life. Some displacements by modern engines were facilitated by the upgrading of lines and strengthening of weak bridges.

Value for money was also obtained from some of the older types of GWR 0-6-0s, such as the double-framed Standard, '322' and '360' classes which dated from the 1860s and 1870s. Many of these engines were eliminated just before Grouping, having been replaced by new 2-6-0s, but '322' class No 354 spent its last 12 years at Leamington shed until 1934, having outlived many later engines. It was not uncommon on the Big Four for the last survivor of a class apparently to escape withdrawal although its class-mates had gone years before.

Leamington Spa was as good a place as any on the GWR from which to observe the changes in motive power through the 1920s. The town lay on the

Above:
A long-lived engine - as late as 1930 Armstrong Beyer Peacock-built '322' class 0-6-0 No 354 was photographed on Lapworth water-troughs, between Birmingham and Leamington. Built in 1866, it spent the last 12 years of its life at Leamington shed.
H. Gordon Tidey T6201

Northern main line from Oxford though Banbury to Birmingham Snow Hill which had been built as the Birmingham & Oxford Junction Railway. The first section of the B&O opened in 1850 from Oxford to Banbury initially as a broad gauge single line. The rest of the route to Birmingham was completed as a double-track mixed gauge line two years later.

Most people who have arrived by train at Leamington from the south will recall a lengthy viaduct taking the railway across the rooftops of the town. This feature was a result of a change of plan before construction. There were ideas that the new main route and the Rugby & Leamington Railway's line between those places might share a stretch of line but no agreement could be made. Instead each company built its own right of way and accordingly Leamington suffered from an unnecessary duplication of embankments and bridges, as well as separate stations for each railway. The London & North Western station was known as Leamington (Avenue).

In its early days, the GWR had two stations at Leamington - one which had broad gauge tracks only and a Brunel overall roof, the other served by mixed gauge lines. The former broad gauge station remained, with the main building at rail level and approached from road level by ramps. In time, the overall roof was cut back and, in common with some other GWR stations - notably Oxford and Banbury -

by the 1930s the station was both depressing in appearance and rundown in condition. Leamington Spa station was reconstructed in 1938 and provided with neo-Georgian buildings which made a minor concession to modern fashion by the inclusion of Art Deco fixtures and fittings.

And what of the services using Leamington GWR? The GWR's main line to the North was just that, for in its timetable headings the railway spoke of services between 'London, Oxford, Birmingham, Wolverhampton, Shrewsbury, Chester, Liverpool and Manchester'. From 1910 the shorter through route via Bicester had replaced the previous trek by way of Oxford and reduced the mileage between Paddington and Wolverhampton by nearly 19 miles. The principal expresses were advertised as Paddington-Birkenhead

Below:
'Star' 4-6-0 No 4024 *Dutch Monarch* takes water from Ruislip water-troughs when working an up Birmingham line express in the late 1920s. Its train includes some recently-built bow-ended corridor stock. *LPC/Ian Allan Library*

Woodside but generally the main section of the train ran as far as Wolverhampton Low Level only. From the 1920s there was a through service in each direction - not always daily in earlier years - between Paddington and the Cambrian Coast. Although the principal London expresses travelled via Bicester, there remained two or three trains that were routed through Oxford, including an overnight working. In addition, there were the Birkenhead-Bournemouth, Birkenhead-Dover and summertime only Wolverhampton-Hastings expresses in each direction. Leamington was the terminus of suburban trains from Birmingham Snow Hill and the remaining GWR passenger trains serving the town included stopping services to and from Banbury and Oxford.

The start from the platform at Leamington was taxing for heavy southbound expresses. GWR 4-4-0s had been replaced on the principal trains before World War 1 by 'Saint' 4-6-0s which later gave way to 'Stars', then 'Castles' and, from 1928, 'Kings' became the motive power on heavier workings. The business trains in the morning and evening apart, many of the Birmingham line expresses of the 1920s were surprisingly light, and often of no more than six coaches. Leamington was served by most of the principal expresses and also by a single slip carriage working.

The bystander on the platform at Leamington would have observed a number of freight trains. Much was general freight and included overnight fast goods trains between Paddington and Manchester and Paddington and the West Midlands, as well as to and from Birkenhead, including a fully braked train conveying 30-40 wagons of Irish meat to London Smithfield and usually referred to as the 'Meat'! There were cross-country vacuum-fitted freights such as between Wolverhampton and Basingstoke.

Left:
Another scene from Twyford cutting, with a down express hauled by one of the versatile '43xx' 2-6-0s which enjoyed a brief heyday on main line work from Paddington. Its train includes 70ft 'Toplight' corridor coaches. *Ian Allan Library*

Below:
An ex-War Department 2-8-0 of Robinson design, as GWR No 3014. Built by North British in 1919, it was acquired by the GWR in the same year when virtually new. The Westinghouse air pump was later removed and this engine and the 19 others in this batch were brought up to GWR standard by the late 1920s. No 3014 remained in service until 1948. *Ian Allan Library*

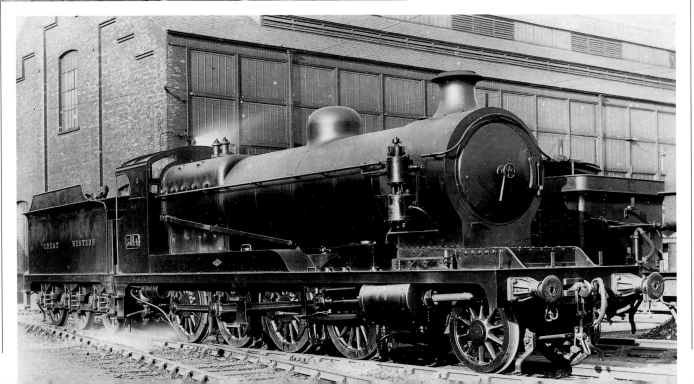

Immediately recognisable were the loaded and empty trains constituting the flow of ironstone traffic from Ardley and the Banbury area to the steelworks at Wrexham and nearby Brymbo.

These ironstone trains were hauled by 'Aberdare' 2-6-0s during the 1920s and by ROD 2-8-0s, both types giving way in time to '28xx' 2-8-0s. These powerful and successful engines numbered only 84 at the start of the 1920s and were not further multiplied until the end of the 1930s. More numerous were the capable '43xx' Moguls of which 322 had been built between 1911 and 1925. The class was well distributed among the GWR Divisions.

The '43xx' and '28xx' classes exemplified Churchward's approach to standardisation. A number of basic components were standardised and employed for the design of new types required by the traffic departments. GWR locomotive engineer Harold Holcroft explained that when he suggested to Churchward that a 2-6-0 might be useful as a general purpose locomotive all over the system, Churchward's instructions were: 'Very well then; get me out a 2-6-0 with 5ft 8in wheels, outside cylinders, the No 4 boiler and bring in all the standard details you can.' That was the Swindon approach to locomotive design and one that was to serve the Company well for 30 years. The 2-6-0 was effectively a tender engine version of the existing '3150' 2-6-2T and made use of the 'Saint' class cab; this did not prove suitable in service and the footplate was instead extended to accommodate the 'County' class cab.

Yet for all the excellent standard Churchward classes, a number of stalwart GWR engines survived through the 1920s. These included the 'Metropolitan' 2-4-0Ts (built 1869-99) on London suburban trains, the '517' 0-4-2Ts and '3571' 0-4-2Ts on country branch lines, and the '36xx' 2-4-2Ts dating from 1900-3 on suburban stopping trains. Already Churchward standard designs had appeared for working some of the local and stopping train duties. There were also the Churchward 'County' tanks previously mentioned which began to be withdrawn from the early 1930s. Churchward had produced a large and small design of 2-6-2T, the prototypes of which had appeared in 1903 - Large 2-6-2T No 99 - and 1904 - Small 2-6-2T No 115. These had been the introduction to the building of the '44xx' and '45xx' Small Prairies (although the GWR did not make much use of this term until much later) and '31xx' Large Prairies. By the first year of World War 1 there were 40 Large Prairies and 66 Small Prairies; it was not until the late 1920s that both types were constructed in enough numbers to make inroads into older tank engine classes. The '5101' class superseded the earlier '31xx' design.

Comparative backwaters such as the Worcester Division made use of the older engine classes for much longer. Although Worcester was busy as a railway centre, the city grew slowly through the 20th century and there was little opportunity for suburban services to develop. There were however stopping trains along the former Oxford, Wolverhampton & Worcester line towards Stourbridge and Birmingham, and via Dudley to Wolverhampton, on the GWR route towards Evesham and westwards to Malvern and Hereford, and on the LMS to Gloucester and via Bromsgrove to Birmingham. Worcester's main branch line was to Leominster via Bromyard, which on its outer stretch was a rural line if ever there was one. Some trains ran from Worcester via Hartlebury and along the Severn Valley to Shrewsbury.

During the 1920s the fleet of express and principal mixed traffic engines on the GWR began to change markedly in composition. By late 1926 there were 35

Right:
GWR railway centre - Worcester Shrub Hill station. This is actually an early 1930s view with 'Bulldog' 4-4-0 No 3362 *Albert Brassey* leaving with a stopping train.
LPC/Ian Allan Library

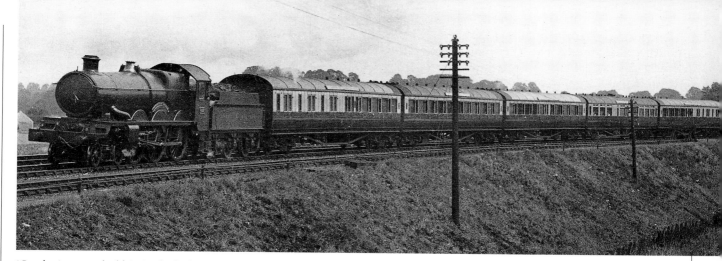

'Castles' at work (this included engines converted from 'Stars') and they were employed particularly on the heavy Paddington-West of England trains and displaced 'Stars'. In turn, this class gained express work from the two-cylinder 'Saints' although many GWR enginemen preferred the latter.

Then, in 1927, came the mighty 'Kings'. Justifying their construction, the GWR explained that over the previous 25 years by strengthening bridges and improving track it had been steadily upgrading its main lines to take more powerful locomotives. The loading of trains had increased and the case had been made for a more powerful design than the 'Castles'. The particular value of the 'Kings' was that schedules of key trains such as the 'Cornish Riviera' could be cut because, as the GWR explained in its various publications for enthusiasts, the 'Kings' could maintain a higher speed upgrade.

It may not be a popular observation, but the cleanliness of the line's engines was not all it could be. Many engines never looked sparklingly clean, and

Above:
A fine panoramic study of the down 'Cornish Riviera Express' passing Newbury in 1927. The engine is one of the first series of 'Castles' with the smaller 3,500-gal capacity tenders. The train includes a mixture of bow-ended and 'Toplight' 70ft coaches. *Ian Allan Library*

Below:
A fine study by a GWR official photographer of a Falmouth-Paddington express at Burlescombe, on the western approach to Whiteball summit on the Exeter-Taunton section of the West of England main line. In a photograph dated July 1929, the engine is No 6018 *King Henry VI* which had entered traffic the previous year. Its train comprises bow-ended corridor stock, including a triplet restaurant car set. All coaches appear to be in the plain chocolate and cream livery introduced in 1927. *Ian Allan Library*

some inside observers have said that after the early 1920s the GWR reduced the quality of the green paint used. One exception was No 6000 *King George V* which in photographs of the time so often appears beautifully turned out.

Through the late 1920s the Big Four companies cut back on standards in the face of lost revenue, largely a consequence of the General Strike and depressed economic activity but also reflecting road competition. The GWR's winter 1928/29 timetable brought a daily reduction of 2,500 miles run by passenger trains, with the saving, it was said, of 8½ engines and 13 sets of trainmen. Among the express trains axed was the 2.30pm Cheltenham St James-Paddington, otherwise known by the British public as the 'Cheltenham Flyer' express, or the 'Fastest Train in the British Isles'. The express was restored to the timetable in the summer of 1929.

On the GWR, the cost savings included simplification of the coaching stock livery, with the elimination of lining-out as from mid-1927. From 1923, the GWR had built numerous bow-ended corridor and non-corridor stock. These coaches were of less expensive construction than the handsome 'Toplight' stock associated with the pre-1914 GWR and although they lasted well enough they were hardly inspired either in internal furnishings or decor. The corridor stock was of a 57ft length until 1929 when a 60ft bow-ended vehicle began to appear. Among the bow-ended stock were articulated corridor and non-corridor sets whose individual vehicles were shorter than the standard 57ft. The articulated corridor stock included some triplet restaurant car sets which were used for a time on crack expresses such as the 'Torbay Express', and to Wolverhampton on the 6.10pm from Paddington, the so-called Birkenhead Diner.

The SR in the 1920s

Viewed from the late 1940s, the Southern Railway appeared to be the best integrated of the Big Four companies but at Grouping the Company had faced as daunting a set of challenges as the other three undertakings. For a start, the South Eastern & Chatham (SE&CR) was a working union only, rather than a unified railway, and there was a sense of injustice on the part of the Chatham shareholders who believed they had received less than their due. The South Eastern, Chatham and Brighton interests were ranged against the London & South Western who they feared would dominate the grouped company. There was a difference of opinion on electrification policy because the Brighton employed a 6,600V ac overhead system while the South Eastern & Chatham proposed 3,000V dc suburban electrification with third and fourth rail. The SE&CR Board actually resolved to proceed with this system in the autumn before Grouping took place but the decision to standardise on the South Western's 750V dc third-rail electrification superseded this unhelpful show of unilateralism.

Yet, so far as management was concerned, the South Western influence was to prevail and from that Company Sir Herbert Walker was appointed as general manager of the SR and became one of Britain's really great senior railway officers. The lead-up to Grouping had required much work on establishing a sound financial structure for the new Southern Railway and the strength of the Company in that respect was one of

Below:
The triumphal entrance to Dover Marine station that was the pride of the SE&CR, whose initials appear over the arch. Waiting to take up an afternoon boat train for Victoria is 'Lord Nelson' No E853. *Sir Richard Grenville. H. C. Casserley*

its great achievements. It needed to be because in its first 10 years the Southern spent a total of £25 million (£1,250 million at today's prices) on improvements and renewals. Much of this expenditure was not immediately apparent because it involved re-equipping the routes to the Channel ports, the facilities at the ports themselves and renewing the ships, all of which had borne the brunt of wartime military traffic to the Western Front and other war zones.

Then there was electrification of the suburban lines up to 1932 which took some 40% of investment. From 1923 to New Year's Day, 1933 with the inauguration of electric working to Brighton, the mileage of electrified lines increased from 240 to 979.

The unification of the three main companies by the Southern occupied much effort. The building of locomotives was confined to Eastleigh and Ashford (Brighton was brought back into the frame in 1943), the bodies of coaches were manufactured at Eastleigh only, while Ashford built all wagons, and the former London, Brighton & South Coast Railway (LBSCR) works at Lancing was used for the manufacture of coach underframes and for coach overhauls. London goods depots were amalgamated. Apart from the improvements to stations and ports, locomotive depots were modernised and carriage sidings improved. The Southern also invested in new signalling installations including colour-light signalling and power signal frames and associated track circuiting.

Above all, the Southern worked hard to modernise its fleets of locomotives and rolling stock and make them more efficient. At Grouping, there were on average 167 locomotive failures a month but five years later the total

Above:
One of the SR's station rebuildings was at Hastings which was opened in July 1931. *Ian Allan Library*

Below:
In preparation for the electrification of former SE&CR suburban services, the track layout and signalling on the approaches to Cannon Street station, London were remodelled in 1926, including the provision of these four-aspect 'cluster'-type colour-light signals. The soon to be removed semaphore signals appear in the background. *Ian Allan Library*

The former London & South Western Railway locomotive depot at Nine Elms, London, as seen in early SR years. Taking water is an ex-LSWR 'N15' 4-6-0 while somewhat ironically alongside the coaling stage is ex-LSWR Drummond 'L12' 4-4-0 No 420, which is running as an oil-burner at the time of the General Strike. *Ian Allan Library*

Inset:
Long-lived Beattie 2-4-0WT SR No 0314 alongside china clay dries in Cornwall. The engine has a Drummond boiler but Adams-type stovepipe chimney. It was renumbered in the 3xxx series after 1931. *Ian Allan Library*

had dropped to 75. More than 1,000 excellent new corridor coaches were built from 1923 to the early 1930s for re-equipment of most main line services and 40% of the wagon fleet was replaced. The Company did not build any new non-corridor coaches for steam haulage and thriftily rebuilt existing stock for local services, many being for push-pull operation.

When it came to steam locomotives some old relics had been inherited and many of these continued at work for some time. The almost immortal Beattie 2-4-0 well tanks remained at work from Wadebridge and were to outlive the Southern Railway. The Adams 4-4-0s had once numbered nearly 120 in two groups, built 1879-87 and 1890-96 respectively. Some of these were eliminated in the first few years of Grouping but

the later engines put in at least another decade of service and a handful survived until the 1940s. No 563 of the 'T3' class was fortunately preserved and restored by the Southern Railway.

Many of the LSWR engines built to the designs of Messrs Adams and Drummond gave sterling service over 50-70 years. While the Drummond 4-4-0s of classes 'L11' and 'L12', for instance, had been withdrawn by the early 1950s and replaced by more modern steam locomotives, classes such as the 'T9' 4-4-0s, '700' class 0-6-0s and 'M7' 0-4-4Ts remained until dieselisation either removed their duties, or else

Right:
Ex-LSWR Adams 'X6' 4-4-0 No 662 was photographed near Okehampton in August 1928, on a stopping train that includes through corridor coaches. No 662 was built in 1895 and withdrawn in 1933. *H. C. Casserley*

Below:
The 'Atlantic Coast Express' was so named in July 1926. In winter, at that time, the train comprised 10 coaches which served Ilfracombe, Torrington, Plymouth, Padstow, Bude, Exmouth and Sidmouth, the restaurant cars being detached at Exeter Central. This picture shows the down train forging along the four-track section between Woking and Basingstoke behind the now preserved 'King Arthur' 4-6-0 No 777 *Sir Lamiel*, its train being made up largely of Maunsell corridor stock.
C. Brown/Ian Allan Library

they were replaced by LMS-design or BR Standard engines made redundant by dieselisation. While the early Drummond 4-6-0s were far from successful and had been thankfully withdrawn or reconstructed when convenient, the Urie 'N15' 4-6-0s dating from 1918-23 proved solid, long-lasting locomotives. They were modified by the SR and named as 'King Arthurs', and continued at work until the mid/late 1950s.

On the South Eastern & Chatham Railway, the maximum 17½-ton axleload permitted on the former Chatham main line prohibited the use of large 4-4-0s and 4-6-0s and, in any case, the SE&CR could not afford to build suitable new engines. Instead, 'D' and 'E' class 4-4-0s were extensively rebuilt with new boilers and new cylinders with piston valves in place of their original slide valves. The result was an extremely efficient and capable engine. Between 1917 and 1927, 32 'D1s' and 'E1s' were produced, for use on the Chatham main line on Kent Coast expresses, and for boat trains between Victoria and Folkestone Harbour and Dover Marine. It was 1934 before 4-6-0s were passed to work between Faversham and Ramsgate so the 'Rebuilds' - as the 'D1s' and 'E1s' rebuilt by Maunsell were known - had a reasonable innings on the most important Kent Coast expresses.

The 'Rebuilds' were only one example of the capable engines the Southern had inherited from the SE&CR. Others included the first Maunsell Moguls of classes 'N' and 'N1', later multiplied and developed, and the Wainwright 'C' 0-6-0s (numbering 109) and 'H' 0-4-4Ts (66 in all).

Wainwright had rebuilt the Stirling 'R' class 0-6-0 shunting tanks which dated from 1888-98 with domed boilers, to become class 'R1'. These engines were used for shunting in Kent, on the Canterbury & Whitstable line and for banking boat trains on the steeply graded Folkestone Harbour branch. Similar reboilering of Stirling 0-4-4Ts produced the 'Q1' class while the Kirtley 'R' 0-4-4Ts built for the London, Chatham & Dover Railway were fitted with the boilers used for the 'H' class; some were later equipped to work push-pull trains. Mention must be made of the diminutive Wainwright 'P' 0-6-0Ts, eight of which were built in 1909/10 for light branch and push-pull trains but soon redeployed as shed pilots and for carriage shunting.

Opposite Top:
No 787, one of the 'L1' 4-4-0s built for the Southern Railway by North British Loco in 1926. *Ian Allan Library*

Opposite Left:
Dover Priory station of the former London, Chatham & Dover Railway in 1931, in the process of being remodelled by the SR. *Ian Allan Library*

Before World War 1 the SE&CR had ordered 22 'L' 4-4-0s from British and German builders for work on the South Eastern main line which had been strengthened to take heavier locomotives. Unfortunately, the 'L's proved to be underpowered for working the fast Charing Cross-Folkestone line expresses and so in 1926 the Southern Railway ordered the modified 'L1' engines, of which 15 were constructed by North British Locomotive Co. The 'L1s' met the specification and had a relatively brief heyday on front-line express work until the early 1930s.

While the Southern Railway constructed numerous up-to-date corridor coaches for work on the Eastern Section (the lines of the former SE&CR), from the Edwardian years the SE&CR had built some attractive corridor coaches. These included non-, semi- and corridor coaches featuring 'Birdcages' at the outer ends of brake coaches.

'Birdcages' (officially called 'roof observatories'!) were semi-circular, raised projections with a view front and aft. They ceased to be a feature of new SE&CR stock after 1915 and were absent from the three sets of Continental boat train stock built in 1921-23 (and later multiplied by the SR). The 'Continental' coaches were most distinctive and compared favourably to any contemporary corridor stock. They were constructed to a length of 64ft, had end vestibule doors only rather than doors on the corridor side opposite each compartment, matchboard panelling on the lower body sides, Pullman-type gangways and couplers, and an absence of gangways on the outer ends of the brakes.

At the time that the 'Continental' stock went into service, ordinary expresses on both the Chatham and South Eastern main lines were composed of non-corridor stock. In 1924/25, the SR put into service several sets of new corridor stock (sometimes referred to as 'Thanet' stock) for these workings but, unlike the 'Continental' stock, they were fitted with British Standard gangways despite the fact that they worked with Pullman cars.

The older stock passing to the Southern from the SE&CR included some pretty dreadful gaslit, non-corridor, non-bogie coaches used for the London suburban services. In the early days of the Southern Railway many of the bodies of four- and six-wheeled coaches built in the late 1890s for the South Eastern Railway were utilised for new bogie vehicles in electric multiple-units.

Mention has already been made of the boat train routes to the Channel ports. Folkestone Junction station was a good location for watching trains and it was easy to be surprised by the racket put up by the 'R1' 0-6-0Ts as, out of immediate sight, they stormed the climb up the harbour branch. In the nearby

sidings would be the main line engine waiting to take on the boat train to London; by 1925, 'King Arthurs' were at work on these trains. Between Folkestone and Dover the railway passes along the coast line through tunnels (Martello, Abbotscliff and Shakespeare Cliff) cut through the chalk, a hazardous location as in 1915 - and not for the first time - two miles of railway had been distorted by cliff-falls which temporarily closed this section until 1919.

At Dover, the Southern Railway carried out a number of improvements soon after Grouping, including rebuilding Priory station and demolishing the London Chatham & Dover Railway's Harbour station, improving rail access to Marine station, providing a new goods yard and relocating the locomotive depot. One feature that appears in so many photographs of trains at Dover is the South Eastern Railway's Lord Warden Hotel. There was no

more impressive sight than the 'Golden Arrow' all-Pullman express on its approach to Dover Marine behind a 'Lord Nelson' 4-6-0. These engines were introduced in 1926 when *Lord Nelson* itself was trumpeted by the Southern as the most powerful express locomotive in the country. More of the class were built in 1928/29 and the class then totalled 16. The 'Golden Arrow' took up its proud title in May 1929 and this marked the debut of the SR's latest, fast and initially first-class only steamship *Canterbury*. With the Wall Street crash that same autumn, the premium cross-Channel rail/sea traffic dropped accordingly and the 'Golden Arrow' became less exclusive.

The Somerset & Dorset in the 1920s

Snaking through the two counties on its way to the coast at Bournemouth, the Somerset & Dorset (S&D) was one of Britain's joint railways. In the case of the Somerset & Dorset Joint Railway, the partnership between the SR and LMS was 50:50. Day-to-day operation was in the hands of officers based locally, and at Waterloo and Derby, and they reported to a committee composed of an equal number of SR and LMS directors. The effective independence of the S&D ceased at the close of the 1920s as Parliamentary powers were obtained for the undertaking to be vested in the SR and LMS, a change made effective from 1 July 1930.

To the enthusiast most of Britain's joint railways were considered picturesque survivals. Their position had been criticised however by the Royal Commission on Transport of 1928 which deemed them an 'inconvenient survival, entailing expense in separate administration and accounting' and recommended that they should be merged with one or other parent company. In the case of the S&D, from 1930 the LMS absorbed the locomotive stock and also operated the system on behalf of the joint committee while the SR took responsibility for all maintenance and civil engineering. The change coincided with the S&D's declining fortunes as 1928 had seen a shortfall of nearly £20,000 between expenditure and receipts and the railway was suffering from road competition.

Much of the attraction of the S&D in the 1920s had rested with its elderly tank engines and curiosities but the locomotive stock handed over to the LMS included many recently-built engines to standard Derby designs. In 1928, the S&D's locomotives and rolling stock comprised 14 4-4-0s, 11 2-8-0s, 26 0-6-0s, 15 0-6-0Ts, 12 0-4-4Ts and two 0-4-0Ts, with 159 coaching stock vehicles, 58 wagons and 171 service vehicles. By now all engines were painted black, the S&D's attractive dark-blue livery having been discontinued although the coaching stock had remained blue until at least 1930.

The works at Highbridge which had been presided over by a resident locomotive superintendent was closed as a result of the 1930 changes although some of its wrecked buildings survived into the 1960s. Works staff were dispersed: some transferred to Norfolk to the Midland & Great Northern Joint (MG&N) line, only to return to the West Country in 1936 when the LNER took over the running of that undertaking. Although coming under direct control of the main line partners after 1930, the S&D only gradually lost its more obvious individuality and many operating practices survived unchanged for a further 30 years.

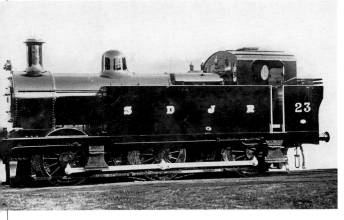

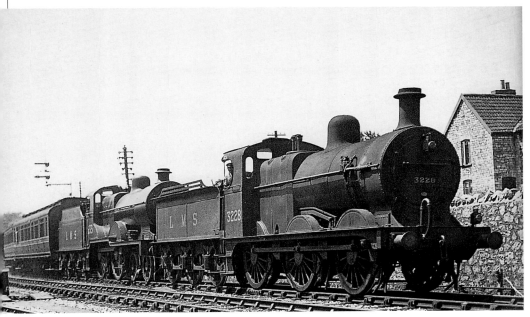

Above left:
Somerset & Dorset Joint: S&D No 23 was one of seven 0-6-0Ts built in 1929 by Bagnall, but soon transferred to LMS stock as No 7154, and later No 7314.
Ian Allan Library

Left:
Somerset & Dorset Joint: the 'Pines Express' was the through express service between Manchester/Liverpool and Bournemouth. It is seen on 7 June 1930, passing Radstock behind two S&D engines which by now have been incorporated in LMS stock. 0-6-0 No 3228 (S&D No 74 of 1902) is leading 4-4-0 No 633 which was built at Derby in 1928 as S&D No 44. *H. C. Casserley*

The LMS in the 1920s

Moving from the S&D with its 105½-mile total mileage (less than 50% of which was double-track), by contrast the LMS owned 6,778 miles. LMS trains also ran over partly owned metals, joint lines and sections of lines over which the Company enjoyed running powers. With over 19,000 single-track miles, the LMS was the most extensive of the Big Four railways and operated the most locomotives and rolling stock. Although the Company operated under the title of LMS as from New Year's Day 1923, the Lancashire & Yorkshire (L&YR) had amalgamated with the London & North Western (LNWR) exactly a year earlier. Neither the North Staffordshire nor the Caledonian Railway had completed their amalgamation arrangements on 1 January 1923 and the process was concluded over the next 12 months.

Operated somewhat imperiously from headquarters offices at Euston station, the LMS had a senior management of officers who were among the toughest and most ruthless of the Big Four. Centralised control and a sharp eye for possible economies were their guiding principles, not that these were necessarily negative attributes in themselves, and in many of their approaches they tackled problems in the manner of modern business corporations.

Policies in Scotland attracted criticism from railway enthusiasts of the 1920s as the bright hues of pre-Grouping liveries were snuffed out by unrelieved and usually grimy black paintwork although immediately after Grouping some engines had been accorded LMS crimson-lake livery. Adornments such as the shapely smokebox wing-plates on Caledonian engines were also removed and pop safety valves fitted in place of the more attractive Ramsbottom design. Major reductions were made in the stocks of some pre-Grouping engines, particularly those of the Glasgow & South Western Railway (G&SWR). Railway photographers of the 1920s recorded an amazing diversity of types, particularly at stations such as the G&SWR's St Enoch, Glasgow, where several classes of 4-4-0s could be seen on local and suburban trains, elegant if none-too-robust 4-6-0s headed Anglo-Scottish expresses (until LMS Compounds began to appear in the later 1920s) and the Whitelegg 4-6-4Ts may have seemed imposing but their appearance deceived for they were easily outclassed by the LMS 2-6-4Ts that later replaced them.

In pursuit of its policy of rationalising the Scottish locomotive stock, the most numerous classes, principally those of Caledonian design, were transferred to former G&SWR lines and to the Highland section. Indeed, after Grouping and until 1925, the LMS had continued to add locomotives of Caley design to stock, including Pickersgill '60' class 4-6-0s and 10 0-4-4Ts basically of McIntosh design. Where possible, standard boilers were fitted to engines seen as having a potential life of more than 10 years or so.

On the Highland section, largely because of the restrictions on axleloading, there was little scope during the 1920s to replace the HR engine types which included a number of attractive Jones 4-4-0s ('Skye Bogies', 'Straths' and 'Lochs'), 4-4-0Ts and diminutive shunting engines. The 4-4-0s especially were best suited to coping with the sharp curves on the Kyle and other lines. From the late 1920s the position changed, commencing with the arrival of Hughes 2-6-0s for main line work, and a few of each of the LMS standard '4F' 0-6-0s and '3F' 0-6-0Ts. The engine type which changed the picture completely on Highland lines was the Stanier '5' 4-6-0 which did not arrive until the mid-1930s, as will be explained in Chapter 3.

Some Midland Railway classes were numerous - one reason why Midland designs were adopted by the newly formed LMS. Some have termed the process 'Midlandisation' to describe the Company's choice of the Compound 4-4-0, the superheated class '2' 4-4-0, the Class '4' 0-6-0 and the '3F' 0-6-0T as the basis of LMS standard types. A total of 1,320-odd of these four types was produced from the early years of Grouping.

There were of course huge numbers of some types of Midland Railway engines. Out of a total of 2,925 engines passing to the LMS, no less than 54% were goods tender engines, made up of no fewer than 471 double-framed 0-6-0s of Kirtley origin, 453 Class '2' Johnson 0-6-0s, 482 Class '3' large boiler Johnson and Deeley 0-6-0s and 192 Class '4', the last-named being the so-called 'Big Goods' engines which were the only really recent members of the fleet. The rest of the Midland locomotive stock was dominated by 226 Class '1' 0-4-4Ts, 142 superheated Class '2' 4-4-0s and no fewer than 245 Kirtley and Johnson 2-4-0s. The majority of the stock was aged and some such as the 0-4-4Ts and 2-4-0 tender engines were obsolete. Around 80 or so of the Midland 0-6-0s had been used to support British forces in France during World War 1.

The strength of these Midland types was that they were strongly constructed, simple and reasonably accessible for servicing. One weakness was that they were often underpowered and hence double-heading

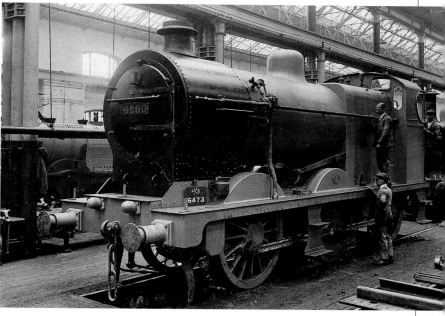

Right:
'Midlandisation' - LMS Standard '4F' 0-6-0 No 4260 in the final stages of construction at Derby Works during the summer of 1926. This engine remained at work until the autumn of 1964.
Ian Allan Library

Below:
The 114-bedroomed golf hotel at Gleneagles, Auchterarder, Perthshire was planned by the Caledonian Railway but its completion was delayed by World War 1, and it opened under LMS control in 1924. *Ian Allan Library*

was frequent, particularly on major flows of coal traffic, something that became even less acceptable once labour costs increased after 1919. Another defect was that aspects of their mechanical design were unsatisfactory - they were prone to running hot-boxes on the coupled axles because the bearings were undersized, the front end design was poor so that the free flow of steam into and out of the cylinders was hampered, engine springs were not easily adjustable in situ and the layout of pipework in and around the boiler made it difficult to keep the engines steam-tight. When some of the LMS standard designs percolated north of the Border, enginemen were surprised at the antiquated nature of some of their features.

Apart from suburban trains into St Pancras originating at Luton and St Albans, there was a frequent service between Barking and along the Tottenham & Forest Gate Railway to Kentish Town and Moorgate. Some trains on the Barking line ran to and from Southend. Principal motive power for these workings were the Johnson 0-4-4Ts, assisted by a few of the Kirtley outside-framed 0-4-4Ts. On the Midland main line suburban trains, although the Johnson 0-4-4Ts featured, also at work were the Deeley 0-6-4Ts (nicknamed 'Flat-Irons' on account of the design of side tanks with cut-outs), 2-4-0s, and even some of the elegant 4-2-2s which were approaching the end of their lives. Then came some experimentation with ex-London Tilbury & Southend Railway (LT&SR) 4-4-2Ts and the same railway's imposing 4-6-4Ts, and even North Staffordshire and Caledonian tank engines were tried, none with any great success. By 1929/30, the competent and handsome Fowler 2-6-4Ts were at work on these trains.

Midland express passenger motive power included the rebuilt and superheated '2' 4-4-0s. These came from various batches but in their most familiar form they had been rebuilt from 1909 onwards and carried either saturated or superheated G7-type Belpaire boilers, these engines carrying numbers in the series 328-562. At first sight the Midland apparently produced few new passenger engines in the years before and after World War 1, but, from 1911 and until Grouping, no less than 157 Class '2' 4-4-0s were subjected to costly rebuilding which included new boilers, and new main frames and cylinders. There were of course the Midland Compounds which by early 1909 numbered 45, some saturated, others superheated.

Below:
An interesting line-up of engines at the Plaistow depot of the former London Tilbury & Southend Railway. From right to left the engines are: ex-LT&SR 0-6-0 No 22899 (built for Turkey, but purchased instead by the LT&SR), LMS Standard '3F' 0-6-0T No 16580, ex-Midland Railway Johnson '1P' 0-4-4T No 1290, and 4-4-2Ts Nos 2119 and 2125, both built by the LMS but to basic pre-Grouping design. *E. R. Wethersett*

In late 1923, the LMS organised comparative trials over the Settle & Carlisle line, pitting a Compound against an ex-London & North Western Railway (LNWR) 'Prince of Wales' 4-6-0 and a Midland '999' 4-4-0, the outcome being that the Compound was demonstrably the most economical of the trio. A further set of trials conducted over the following year involved an LNWR 'Claughton' 4-6-0, a Caledonian 4-4-0 and three Compounds; again, the Compound proved the most economical. In view of its modest size, it did well to hold its own against the 4-6-0 in terms of power output.

The results of the 1924/25 trials vindicated the decision taken by the LMS to select the Midland Compound as the basis for its principal express engine although the post-Grouping engines (numbered from 1045 upwards, with later engines from 900 upwards) had 6ft 9in driving wheels. Apart from their use on former Midland and LNWR main lines, the Compounds were also allocated to Northern Division sheds (to the LMS, Scotland was its Northern Division).

Speaking of the Midland lines, Derby was a natural centre as it was here that the West of England and St Pancras-Sheffield-Leeds lines met. The East Midlands town was the home of two major workshops - Derby Locomotive Works and the carriage works at Litchurch Lane - and extensive headquarters offices. The West of England main line featured a number of through expresses between Bristol and Leeds, and in summer through services to the South Coast at Bournemouth by way of Bath and the Somerset & Dorset line.

Midland Division through services reached both Bristol and Bournemouth, and away to the east through trains from St Pancras served Southend-on-Sea. The Midland had taken over the former London, Tilbury & Southend Railway in 1912 and its services were largely left unchanged, the principal London terminus for its services being the Great Eastern Railway's Fenchurch Street station in the City. The Midland had undertaken to electrify the system but such plans were aborted with the outbreak of World War 1.

The LT&SR 4-4-2Ts were the mainstay of an intensive suburban service via Upminster and via Tilbury. The Midland's main contributions were to tinker with the design of the 4-4-2Ts and to downgrade the grade of coal used such that reliability suffered. In contrast with the Midland, the LT&SR used air braking. The LT&SR engine fleet passing to the LMS also included some 0-6-2Ts, and a pair of 0-6-0s originally ordered for service in Turkey. There were eight Whitelegg 4-6-4Ts which were banned from working from Barking into Fenchurch Street and so came to be redeployed elsewhere. The major running depot was Plaistow.

Another formerly independent railway in the London area was the North London Railway (NLR) which had been worked by the LNWR from 1909. Services from Broad Street to Richmond and Watford had been electrified from 1916 and 1922 respectively but there remained steam-worked local passenger trains into the East End, as well as a heavy goods service in the London docks. The NLR stock of 4-4-0Ts (which had been withdrawn from service by 1925) and outside-cylindered 0-6-0Ts was maintained at Bow Works and operated from the nearby Devons Road running shed. One peculiarity was an 0-4-2 crane tank nominally dating from 1858 and usually employed within Bow Works.

The 100 or so North London engines were just a drop in the ocean when it came to the 3,200 LNWR engines. The Company's passenger tender engines numbered 896 on the last day of 1922 and included 2-4-0s of 'Precedent' and 'Waterloo' classes, Webb 4-4-0s of the 'Alfred the Great' and 'Renown' classes, the fine 'Precursor' and 'George V' 4-4-0s, and three classes of 4-6-0, the 'Experiments', 'Prince of Wales' and 'Claughtons'. Many of the last two classes had been built during or after World War 1.

There was no doubting the capability of the more recent of the LNWR express engines when in good condition. Some inherent design faults, combined with the fact that the engines had been manufactured as cheaply as possible, meant that their mechanical state deteriorated both with mileage and with reduced maintenance after World War 1. Some of their mechanical parts were flimsily built, the specification of some of the components was not up to what was demanded of them and, in common with the Midland engines, the bearing surfaces were inadequate. The contrast between what the principal ex-LNWR engines were potentially capable of and their run-of-the-mill performance was marked. In the case of the 'Claughtons', the LMS tried hard to rectify their mechanical defects and attempted to give special attention to those required for the premier workings. There is the impression that by the late 1920s the Company had concluded that it was a matter of throwing good money after bad, and that new engines were the only solution.

That said, many LNWR types lasted longer than one might have expected. Even the 2-4-0s survived into the 1930s, while four 'Waterloos' were transferred to the stock of the District Engineers, named *Engineer Lancaster* or *Watford*, to take two examples, and the last was not withdrawn until the mid-1930s.

The Webb passenger tank engines - 2-4-2Ts and 0-6-2Ts - were hard at work through the 1920s, the former on push-pull workings. The LNWR's rail-motors began to be taken out of service from the late 1920s. As to the larger passenger tanks, there were the Whale 4-4-2Ts and Bowen Cooke 4-6-2Ts. Largest of the

Above:
LNWR 0-8-0 No 2114 approaching Crewe station in early LMS days with a down coal train. *Real Photographs*

LNWR tank engines were the massive Beames 0-8-4Ts, somewhat hampered by their tendency to derail on sharp curves.

The 0-8-4Ts were derived from the Webb 0-8-0s which had a complicated history including those which were built as compounds. The most recent 0-8-0s were of the 'G2' class put into service immediately before Grouping. From 1919 the LNWR obtained on loan a number of 2-8-0s of Robinson design from Government stocks and in due course the best of these were selected for reconditioning and remained at work for the LMS through the 1920s. The 0-8-0s and these 2-8-0s were in a minority as far as LNWR freight motive power was concerned. Five hundred of the Webb 'Coal Engines' had been built, as well as 310 of the designer's 18in Goods, more popularly known as the 'Cauliflowers'. Survivors of both types continued to give good service after Grouping. The tank engine version of the 'Coal Engine' was the Webb 'Coal Tank', of which 291 passed to the LMS. There were various designs of LNWR shunting engines, including the 0-6-0ST Special Tanks, some of which were in departmental service at Crewe and Wolverton Works, and 0-4-0 square tanks as well as the so-called 2-4-0 'Chopper Tanks'.

The way in which the LMS cut down some of the fleets of Scottish pre-Grouping engines has been mentioned. Much the same happened to the engines acquired from the Furness and North Staffordshire railways (NSR), for the same reasons, as both companies owned small classes of modestly powered engines. The smaller railways' workshops were closed down as part of an inevitable rationalisation of facilities. On the former North Staffordshire, Stoke Works had closed by early 1927. During the late 1920s a number of LMS '4F' 0-6-0s were allocated to ex-NSR sheds and then most of the earlier ex-NSR 0-6-0s were consigned to the scrapheap.

The 1,650 locomotives of Lancashire & Yorkshire design inherited by the LMS generally enjoyed a longer life than those of the smaller constituent companies and the Barton Wright, Aspinall and Hughes era 0-6-0s proved both durable and long-lived. So did the Aspinall and Hughes 2-4-2Ts, some of which were transferred to other parts of the LMS. The L&YR also constructed 0-8-0s, of which there were 295 as compared with the 480 North Western engines, but although powerful they proved expensive to maintain.

The L&YR may have amalgamated with the LNWR in 1922 but, taken on its own, it was the third largest constituent of the LMS and became that railway's Central Division, many of its operating practices and train service patterns being left unchanged through the 1920s. Twenty of the Hughes four-cylinder 4-6-0s were built after Grouping and a few of the class were

Left:
The interior of one of the lounge first brakes built for the 'Royal Scot' in 1927.
Ian Allan Library

Left:
Crowds at Birmingham New Street station in 1927.
Science & Society Picture Library DY 14320

employed on West Coast main line expresses north of Crewe where their day in and day out performance proved to be somewhat disappointing.

Apart from the four locomotive types featuring in the LMS's Midlandisation policy, other new engine types built for the LMS in the 1920s included the Horwich-influenced 2-6-0s (known to many enthusiasts as 'Crabs') and notable for being the first entirely new LMS design to appear, in 1926; the Beyer Garratt engines purchased from 1927; the 2-6-4Ts (introduced in 1927) and the 0-8-0s nicknamed 'Austin Sevens' which first appeared in 1929.

After a couple or so years following Grouping it was clear that Compounds were not powerful enough for working the heaviest Western Division and West Coast main line expresses. In 1926, the LMS borrowed the GWR's No 5000 *Launceston Castle* to run trials between Euston and Carlisle and its capability demonstrated the case for a modern 4-6-0 on the LMS. The Company's Board of Directors are said to have wished to see the construction of 50 'Castles' for their railway, but this was impractical for various reasons. Instead, the North British Loco Co was contracted to build 50 three-cylinder 4-6-0s which embodied some of the main characteristics of the GWR 'Castles' although their mechanical design was hardly similar. The new 4-6-0s were the 'Royal Scots', and the first 50 were put into service during 1927.

The 'Royal Scots' made a happy match for the LMS standard corridor stock of the 1920s which generally followed latter-day Midland Railway design, with a mixture of vehicles to side-corridor and open layout, some of which were of all-steel construction. From 1927/8, the wintertime 'Royal Scot' express ran nonstop behind a 'Royal Scot' over the 299 miles between Euston and Carlisle, and the train's formation included no fewer than six vehicles for the service of meals. In winter, the 'Royal Scot' was usually made up to 12 coaches due to be worked unaided by a 'Royal Scot' over Shap and Beattock banks but in summer it was scheduled to consist of 15 coaches of 417 tons.

The LNER in the 1920s

It was to be expected that by comparison with the other Big Four companies the LNER would find it easier to assimilate its principal constituent companies. The Great Northern, North Eastern and North British companies had co-operated closely as the East Coast Conference, and in the operation of the coaching stock forming the East Coast Joint Stock used for Anglo-Scottish passenger trains. Also, the Great Northern, Great Central and Great Eastern Railways had attempted to amalgamate in 1909 and the Parliamentary Bill to facilitate this process had secured a second reading in the House of Commons although it was subsequently withdrawn. So the process of Grouping was likely to be smoother, particularly as the LNER chose to function more as a federation of operating areas, controlled by a conciliatory and effective senior management at headquarters. The Company continued to refer to its individual sections by their pre-Grouping names.

The LNER was never financially secure but within this constraint did its best. The Company was badly affected by the loss of traffic during the General Strike, and in particular by the depressed state of industry in Durham and Northumberland during the mid/late 1920s. The LNER suffered a fall of 13% in passenger receipts when comparing 1923 with 1927, more acute than any of the other Big Four companies.

Given its economic difficulties, the LNER had to defer some desirable modernisation schemes such as electrification of the Great Northern suburban services but the much-needed re-equipment of suburban rolling stock on the Great Northern and Great Eastern London suburban services was allowed to proceed. Pacifics and new coaching stock were built for use on the East Coast route, the East Coast expresses being the *crème de la crème* as far as the LNER was concerned. The premier service was the 'Flying Scotsman' which from 1928 ran non-stop in summer between King's Cross and Edinburgh.

Two major investment schemes were principally concerned with freight movement and began in 1928. The first was enlargement and re-equipment of the huge marshalling yard at Whitemoor, near March in Cambridgeshire, the other scheme featuring a range of improvements to railway facilities in and around the steel town of Frodingham, Lincolnshire, and including modification of the track layout and new marshalling yards, both involving the provision of new signalling, new goods and locomotive depots and a new station.

On the LNER centres out of the limelight such as Frodingham generated the revenue although the East Coast main line tended to get the most publicity. Frodingham was on part of the former Great Central Railway, best known perhaps for its London Extension south from Sheffield, but essentially a cross-country railway with the main line from Manchester to Sheffield and on to Lincoln and Cleethorpes, branches to Wakefield, Barnsley and other centres, as well as outposts in North Wales and Lancashire. Grimsby, at one time Britain's largest fishing port, was developed by the GCR. Over 180,000 tons of fish were landed in 1922, the majority of it distributed by train. Just before World War 1, the GCR had built a vast port at nearby Immingham.

Coal and mineral class traffics were the lifeblood of the GCR whose lines were well placed to serve the Yorkshire, Nottinghamshire and Derbyshire coalfield which had grown rapidly from 1900. A proportion of the output from this coalfield was exported through Immingham, Hull, Grimsby, Liverpool and via ports served by the Cheshire Lines Committee; all such movements took place for at least part of the way over former GCR metals. Some of this export traffic and heavy flows of coal traffic for internal consumption passed by way of Doncaster, Tuxford,

Opposite top:
The nonstop summertime 'Flying Scotsman' from Edinburgh Waverley to King's Cross nears its destination and passes Finsbury Park in June 1928. The engine is 'A1' Pacific No 2547 *Doncaster*. The coaches are recently built Gresley standard teak-bodied corridor vehicles including a triplet restaurant car, but the leading vehicle is an all-steel full brake van. *F. R. Hebron/ Rail Archive Stephenson*

Opposite below:
All around Whitemoor marshalling yard, near March, are the Cambridgeshire fens. The LNER invested heavily in the re-equipment of this entry point for goods to the Eastern Counties and eastern suburbs of London.
Ian Allan Library

Above:
Two Robinson designs for the Great Central Railway - '8A' class
0-8-0 No 1134 of 1907 and inside-cylindered 4-6-0 of the '1' or
'Sir Sam' class, built 1912/13. *Ian Allan Library*

Below:
One of the first series of Robinson GCR 'Directors', No 5437
Prince George passes over Charwelton water-troughs (between
Rugby and Woodford Halse) with a southbound stopping train.
Ian Allan Library

Colwick (the main Great Northern marshalling yards in Nottingham) and Banbury.

In 1911, J. G. Robinson, the GCR's capable Locomotive Engineer, introduced the first of his Class '8K' 2-8-0s. By the outbreak of World War 1 there were 126 of these engines in service. During that war, the Government decided that the Robinson 2-8-0 was the most suitable heavy goods design for wartime service overseas and 521 were constructed. At Grouping, a total of 148 '8Ks' and the similar '8Ms' passed to the LNER which also decided to purchase some of the War Department engines, eventually acquiring 273 of them. The LNER classified the Robinson 2-8-0s as 'O4' and 'O5'.

The Robinson 2-8-0s, nicknamed 'Tinies', were just one of several GCR goods engine types. Their predecessors were the same designer's '8A' class (LNER 'Q4') outside-cylindered 0-8-0s, 89 of which entered service from 1904. Smaller, though similarly competent, were the 174 engines of Robinson's '9J'

class of 0-6-0s which became LNER Class J11 and were nicknamed 'Pom-Poms'.

J. G. Robinson was responsible for a full range of workmanlike, solidly-built and attractive engines, including three main designs of 4-4-0 (LNER Classes 'D9', 'D10' and 'D11'); Atlantics, both simple and compound expansion types (LNER Classes C4 and C5) and a varied selection of 4-6-0 designs, both inside and outside-cylindered (LNER Classes 'B1'-'B9') which were built from 1901 until 1924. The LNER selected the 'D11' design to fulfil a need for more 4-4-0s in Scotland, and 25 were constructed soon after Grouping with detail differences to the originals, including cut-down boiler mountings and cab to suit the loading gauge.

When the LNER was formed, the Company's directors were faced with appointing a Chief Mechanical Engineer, the only senior technical appointment with authority over the whole railway. Of the constituent companies' locomotive superintendents and engineers, some were at retiring age. In terms of seniority, the directors were bound to offer the post of CME to J. G. Robinson but, at the age of 66, he declined. Instead, H. N. Gresley of the Great Northern Railway was appointed CME and he took up his post in February 1923. It was inevitable that Great Northern/Doncaster practice would prevail for the design of locomotives and rolling stock alike.

Some railway writers have wondered what sort of locomotive designs Robinson might have produced had he accepted the post of CME. Locomotive engineers, even the CME of the LNER, submitted designs to meet requirements set out and agreed by the railway's traffic committee or its locomotive and traffic committee. On the LNER the latter was chaired by the Company's Chairman, for so long the redoubtable William Whitelaw.

The LNER went ahead with ordering more Gresley 'A1' Pacifics. As this design already existed, had he become CME Robinson would hardly have chosen the NER Raven Pacifics in preference. Other existing Gresley designs, such as the 'K3' 2-6-0s, which were built from 1924, would surely have been perpetuated, while decisions taken to build a new 0-6-0 based on an existing North Eastern design (and designed by Darlington drawing office) are unlikely to have changed. In meeting the needs of the Southern Scottish Area, Robinson might easily have decided to adopt his own 'Director' 4-4-0 design, which is what happened under Gresley. Robinson would hardly have continued as CME beyond the age of 70, and during the late 1920s would have faced restraints on locomotive building similar to those experienced by Gresley. It was traditionally the task of the CME to propose which types of life-expired locomotives should be considered for scrapping but that was not to say funds were available to replace them. All the Big Four companies had limited expenditure for new steam locomotives during the late 1920s.

As already mentioned, two locomotive designs put in hand for construction by the newly formed LNER were the existing Gresley 'K3' 2-6-0, which had first appeared in 1920, and the first of the LNER Group Standard classes, an 0-6-0 goods engine built as the 'J38' and 'J39' classes. 'J38s' were for service in Scotland and remained there throughout their existence while the larger wheeled 'J39s' were used all over the LNER. The 'J39s' first appeared in 1926, and by the time construction ceased in 1941 they totalled 289. Another design from the Darlington drawing office was the outside-cylindered 'D49' 4-4-0, 76 of which were built from 1927-35, all allocated to the LNER's North Eastern and Southern Scottish Areas. They were intended in the main for 'second division' passenger work although some saw top-link duties in

their early days. The 'D49' class included engines built new or modified with poppet valves.

Of course, some 'true' Gresley designs emanated from the CME's headquarters office in the 1920s, such as the massive 'P1' 2-8-2s designed at Doncaster to work 100-wagon coal trains south of Peterborough. Gresley was always prepared to consider less orthodox solutions and was supportive of the Sentinel-Cammell and Clayton steam railcars which were ordered from the builders from 1925. They were to prove inadequate both in performance and reliability.

The Great Northern 'C1' Large Atlantics proved splendid performers on the 'West Riding' Pullman car train which succeeded where previous Atlantic-hauled Pullman trains failed to generate custom, notably the Sheffield/Manchester service. Although 12 Gresley 'A1' Pacifics were at work by the end of 1923, for another year or so express services out of King's Cross continued to be dominated by the Large Atlantics. Secondary passenger trains at the southern end of the GNR main line made use of the Ivatt small Atlantics (nicknamed 'Klondykes') or the designer's classes of 4-4-0. Many of the other GNR types such as the 0-6-0 goods engines, 4-4-2Ts, 0-6-2Ts and 0-6-0STs remained the backbone of the motive power allocation throughout the Southern Area of the LNER.

While the majority of the leading pre-Grouping classes on the Great Eastern section lasted beyond the lifespan of the LNER, by 1927 the Company felt it necessary to build new express locomotives: no new express engines had been built since the Grouping. Excellent though they were, the ex-GER 'B12' 4-6-0s needed to be supplemented. The GER Section was limited to a maximum axleloading of 17 tons on each driven axle and so a completely new design of express locomotive needed to be thought out carefully. As a stopgap, 10 new 'B12s' were ordered immediately but entered service later than planned, in 1928; by the end of that year, 10 of the new three-cylinder 'B17' class had also entered service.

The North Eastern Area of the LNER corresponded to the territory of the former North Eastern Railway and that of the Hull & Barnsley (H&B) Railway, which during 1922 had merged with its larger neighbour. The Area included 160 miles of the East Coast main line and in the first years after Grouping the principal

Opposite Top:
One of the Gresley 'B17' 4-6-0s, No 2818 *Wynyard Park*, makes a fine sight near Whittlesford when working the 2.5pm Norwich-Cambridge-Liverpool Street. *E. R. Wethersett*

Opposite:
A Great Eastern engine at work for the LNER: the Audley End-Saffron Walden push-pull train, formed of ex-GER coaches and worked by 'G4' 0-4-4T No 8122. Built at Stratford Works in 1900, No 8122 survived only until 1932. *E. R. Wethersett*

expresses using it continued to be hauled by Worsdell and Raven's Atlantics which numbered 72 in all. They were assisted by the five Raven Pacifics, three of which had been built after Grouping. In service, the Raven engines compared unfavourably with their Gresley counterparts which began to be delivered new to North Eastern sheds from 1924.

Coal and mineral traffic provided the bedrock of the NER. The Railway conveyed a larger tonnage of such traffic than any other pre-Grouping railway, and was the largest dock-owning railway, much of the coal being exported through the Tyne ports, Blyth, Sunderland, Hartlepool and Hull. Consequently, the NER had a massive fleet of goods engines of 0-6-0 (no less than 677 in total) and 0-8-0 designs (215 in three classes) which were little changed during their ownership by the LNER.

Mention of the North British Railway (NBR) immediately conjures up images of Edinburgh Waverley station for many people. It was here that the Reid Atlantics (LNER Classes C10 and C11) were in their element, at the head of expresses to Aberdeen, Perth and over the Waverley Route to Carlisle; at around the time of Grouping they worked briefly between Edinburgh and Newcastle. Otherwise, the Southern Scottish Area made extensive use of 4-4-0s on passenger work, particularly the Reid 'Scotts' which worked most of the Edinburgh-Glasgow expresses and the 'Glens' which were associated with the West Highland line. The 4-4-0s often double-headed on the heavier trains. Many of the 0-6-0s that the LNER inherited from the NBR were life-expired and all but a very few of these had been scrapped by

the late 1920s. In contrast, three classes - the Holmes (LNER 'J36') and the beefy-looking Reid ('J35' and 'J37' classes) - had another 40 years of life ahead of them.

Apart from new LNER engines - notably the 'J38' 0-6-0s and 'D11/2' 4-4-0s - the ex-NBR classes largely remained unchallenged on their home territory during the 1920s although some ex-GNR 'D1' 4-4-0s were transferred to Scotland in 1925; these were regarded as 'cast-offs' and were generally confined to menial tasks. Other engines transferred to the Southern Scottish Area from further afield included ex-North Eastern 'J24' 0-6-0s (from 1924) and ex-Great Eastern Railway 0-6-0Ts which arrived from the late 1920s.

The northernmost lines of the LNER were included in its Northern Scottish Area which comprised the former Great North of Scotland Railway (GNSR) lines based on Aberdeen. The network's relatively modest passenger traffic was exceeded in revenue by a variety of freight traffic - fish, agricultural traffic including grain, potatoes and livestock, and forwardings of whisky from the 43 distilleries that the LNER served on Speyside and elsewhere. The majority of traffic - passenger and goods - was handled by one hundred 4-4-0s of 14 classes and sub-classes. Other GNSR engines which passed to the LNER included nine 0-4-4Ts which were employed on the Aberdeen suburban services, two classes of 0-6-0T, and four diminutive 0-4-2Ts built to shunt in Aberdeen docks. 0-4-2Ts built to shunt in Aberdeen Docks.

Below:
Ex-Great North of Scotland 'D41' 4-4-0 No 6900. The GNSR used 4-4-0s for mixed traffic work, and the 'D41s' continued at work until the late 1940s/early 1950s. *Ian Allan Library*

3. **The 1930s:**

From the Depths of Depression to the Streamliner Age - 'A4s' to Air Raid Precautions

The thirties started with the Big Four on the defensive. For one thing, social changes were at work: through the 1920s all four railways lost profitable first-class traffic, both in quantity and in revenue generated, for the differential between first and third class was now much greater than it had been before 1914.

The Government's appointment of a Royal Commission on Railways in 1913 had been particularly concerned with freight rates and had contributed to conditions set out in the Railways Act, 1921, under which the Big Four had been created. After Grouping the scrutiny continued. A Royal Commission on Transport reported in 1930/1 and the tone of its findings and observations on the railways was generally critical. The Commissioners claimed that there had been 'practically no improvement in locomotive speed during the last 80 years' and that passenger services should be speeded up and made more convenient; fares should be both reduced and revised; and there should be a statutory obligation on the railways to provide all passengers with seats on main line trains.

The Commission seemed to spur the railways to make a wide range of improvements in succeeding years. The Big Four speeded up express trains noticeably from 1932, introduced more cross-country expresses (to meet the criticism that rail services were frequently inconvenient), increased the number of wayside halts, and from 1933 introduced the 1d/per mile long-distance Monthly Return cheap fares. Holiday runabout and season tickets, half-day and evening excursions at low fares were all innovations dating from the early 1930s.

The slide in passenger receipts had already begun in 1929 at the time of the Wall Street crash but freight traffic held up a little longer. Overall, the picture was alarming, for 1930 saw the British railways' net revenue drop from 1929's £45 million to £37.7 million. The fall continued as traditional industries took the force of the Depression, so that the railways' net

Above:
The 1930s Depression hit the LNER badly with the decline of traditional heavy industry in the Northeast. Ex-NER 'A6' 4-6-2T No 687 leaves Darlington with a Saltburn train in October 1931.
W. M. Rogerson/Photomatic 7822/Rail Archive Stephenson

revenue in 1932 was just £26.4 million. Depression
also brought price deflation which emphasised the
haemorrhaging of receipts. Economies were sought
variously, by deferment of locomotive and rolling
stock building programmes and of maintenance of
way and works, and by increasing productivity -
bigger engines hauling heavier trains, old engines not
being replaced and old freight yards being cut out.
Staff wages - and jobs - were cut. Lines were closed to
passenger traffic.

It is difficult to isolate the effect of the economic
depression on the erosion of railway traffic.
Competition from road transport also ate into
passenger and freight carryings, particularly on rural
lines. Then there was the unhappy decline of Britain's
agriculture during the 1930s, itself induced by a slide
in commodity prices, and hastening the drift of
population from the countryside to urban areas.

Although trade began to pick up from 1934
onwards, traffic continued to decline on some parts of
the railway system. To take an example of what was
happening on the rural or semi-rural lines, on the
GWR's Severn Valley line from Hartlebury to
Shrewsbury by way of Bewdley and Bridgnorth, the
last-named station booked just under 16,000

passengers in 1935, but only 13,000 in 1939. The
annual total of parcels forwarded slid from 41,000 in
1935 to 37,000 in the same period, while the goods
tonnage forwarded from Bridgnorth declined from
5,400 to 2,300 tons. The Severn Valley line's fortunes
were reversed immediately before World War 2 as
military stations and depots were set up near the line's
stations and generated new traffic.

Despite the emphasis on economies and cut-backs,
some investment was made in better infrastructure as
the railways took advantage of Government schemes.
There was the remission of Passenger Duty under the
Finance Act of 1929, by which the railways could
spend the 'released' money on improvement schemes,
and the Development (Loan Guarantees and Grants)
Act, 1929, by which the railways received financial
assistance with major infrastructure projects, with the
Government both guaranteeing and paying the

Above:
In 1937, a Liverpool Street-Loughton train passes the site of works for the extension of London Transport's Central Line tube, part of the New Works Programme which got under way in 1935. The engine is ex-GER 'F6' 2-4-2T No 7061 which is hauling ex-GER 54ft suburban stock. *Ian Allan Library*

Right:
Insulated containers used by the Southern Railway for the road movement of meat from Southampton Docks to principal towns in the south of England. The Thornycroft lorry carries the Southern Railway name on the radiator. Picture dated 1931. *Ian Allan Library*

interest on the capital expended on these schemes. Under the provisions of this Act, the GWR rebuilt parts of its Paddington, Bristol Temple Meads and Cardiff General stations, quadrupled lines in the Birmingham suburban and Taunton areas, built the avoiding lines at Frome and Westbury, and improved some of its marshalling yards.

The remission of Passenger Duty provided an impetus for the Southern Railway to complete electrification of its Brighton main line. Under another Government scheme, dating from 1935, the railways could borrow money for carrying out new works. This aided projects as varied as the Portsmouth Direct line electrification of the Southern Railway, and improvements made after 1935 to the operation of the GWR's Barnstaple and Minehead branches.

The biggest single scheme was the so-called New Works Programme, an agreement announced in June 1935 involving the Treasury, the London Passenger Transport Board, the GWR and LNER. It aimed to rid steam working from a clutch of London's suburban lines. Financed by Government-guaranteed loans, the Programme provided for the construction of new tube railway in London and its suburbs; electrification of the LNER Colchester main line as far as Shenfield; extension of the London Transport Central Line in

tube from Liverpool Street to Leyton and thereabouts, with operation of tube trains over LNER suburban lines to Loughton and associated loop lines; extension of the Morden-Edgware line (later known as the Northern Line) from Highgate to East Finchley with tube trains running on LNER lines to Edgware, High Barnet and Alexandra Palace; additional tracks alongside the GWR Birmingham main line as far as Ruislip, for use by Central Line tube trains extended from the existing Ealing line, and other improvements purely on the LT network. The New Works Programme was scheduled for completion by the autumn of 1940 but its fulfilment was stymied by World War 2. One or two of its schemes were abandoned altogether.

Improvements in service and facilities were not restricted to passenger traffic. The LMS invested in a number of improvements to freight services including the development of road/rail containers, and some invisible to the casual observer such as the replanning and modernisation of freight terminals to improve their productivity. This railway worked hard to develop long-distance freight services for merchandise traffic where road transport

was competing strongly. The LMS claimed that 90% of all merchandise traffic was delivered to the consignee the day after collection. Its No 1 Fully Fitted Freight trains had all their vehicles fitted with vacuum brakes and connected to the engine, and were allowed to load up to 45 wagons of goods. Another major LMS achievement was the remodelling and mechanisation of Toton marshalling yard in the East Midlands, a project completed during the spring of 1939. Shunting at Toton was performed by the new LMS diesel-electric shunting locomotives as an indication of future trends in motive power. As with several interwar railway projects, Toton's modernisation was completed under the shadow of approaching war.

The GWR in the 1930s

Much of the GWR's public reputation was built around the image of its premier express locomotives - the 'Kings' and 'Castles' - and of the Company's principal express trains, notably the 'Cornish Riviera', the 'Cheltenham Flyer' and the 'Torbay Limited'.

For 20 years the GWR could claim that the 'Cornish Riviera Express' made the longest nonstop run in the world, over the 225¾ from Paddington to Plymouth, a record that in 1926 passed to the 'Royal Scot'. Yet the train's schedule of a level 4hr to Plymouth remained a fine achievement through the 1920s and 1930s, given the usual trailing load of up to 14 coaches, with an all-up weight of some 530 tons. That load applied only as far as Westbury, however, for the rear two coaches, for Weymouth, were slipped there, the Minehead and Ilfracombe coaches were slipped at Taunton, and the two coaches, for Exeter and Kingsbridge, were similarly slipped outside Exeter St Davids. The Newquay coach was detached during the stop at Plymouth, that for Falmouth at Truro, and the St Ives coach, at St Erth. By the time it reached Penzance the 'Cornish Riviera' comprised just three coaches and a restaurant car.

In 1929, the 'Cornish Riviera' had received new sets of coaches and, in 1935, two really splendid sets of 'Centenary' coaches which added a new dignity to the GWR's flagship. The coaches forming the 1929 and 1935 stock were constructed to the GWR maximum permitted loading gauge of 9ft 7in width. Otherwise, the 'Cornish Riviera' remained relatively unchanged right through the 1930s, with the same 'King' class 4-6-0 at its head. Yet the train's reputation did not dull.

The 'Cheltenham Flyer' was something of an enigma. The GWR created a legend - a coloured luggage label was distributed to the train's passengers as a souvenir from 1932, the Company marketed a jigsaw featuring the train the following year, and the year after published the *Cheltenham Flyer* book. 'During the past decade,' said the book's author, W. G. Chapman, 'the average speed of the "Cheltenham Flyer", now the world's fastest steam train, has actually increased by nearly 10mph, from 61.8mph to 71.3mph.' He was referring not to maximum speeds, but the average speed sustained over the 77 miles between Swindon and Paddington.

With the September 1932 timetable, the 'Flyer' was accelerated to run between Swindon and Paddington in 65 minutes, but on 5 June of that year there occurred the 'record of record' runs when the train had reached Paddington in just 56¾ min from Swindon with No 5006 *Tregenna Castle* hauling five ordinary coaches and a restaurant car. Perhaps 1932 was the high-water mark of the 'Flyer'. By the mid-1930s, attention had turned to the GWR's new flyer, the 'Bristolian', which had been introduced in the railway's centenary year. Before long, the streamliners of the LNER and LMS were attracting all the publicity.

Express passenger trains were the icing on the cake, for the GWR's revenue was based on freight traffic and that had been savaged by the Depression. Particularly so in South Wales where coal shipments from Bristol Channel ports declined from 36.7 million

Below:
The 'Bristolian' approaching Bath c1936 with No 5055, then named *Lydford Castle* but renamed *Earl of Eldon* in 1937. The coaches are mostly the so-called 'Sunshine' side-corridor stock which resembled contemporary LMS designs. However, the fourth vehicle is one of the two 'Quick Lunch Bar' cars of 1934.
Ian Allan Library

Right:
One of the locomotives from the South Wales railways absorbed by the GWR: this is ex-Rhondda & Swansea Bay Railway 0-6-0T No 7 (built by Beyer, Peacock 1889) when running as GWR No 806. The R&SB locomotives were absorbed into GWR stock in 1908. Having been the recipient of a number of Swindon features, No 806 survived in service on its home territory until 1940, afterwards being used as a Swindon Works shunter until going for scrap in 1943. *Ian Allan Library*

Right:
'Saint' 4-6-0 No 2939 *Croome Court* on a typical secondary duty, the summer Saturday 1.25pm Paignton-Swindon-Leicester-Sheffield train, photographed at Clink Road Junction, near Frome, in 1937. The first two coaches are of the GWR 'Excursion' stock, there are two LNER vehicles towards the rear, and a milk tank wagon behind them.
LPC/Ian Allan Library

tons to 19.3 million, the latter being just half the total of the peak year of 1913. The consequence was a ruthless pruning of lines, locomotive stock and staff; by 1936, the GWR's staff in South Wales had been reduced by 5,000 in number. With the loss of industry, the Valleys lost population to emigration so passenger traffic also declined. The GWR put in connections between formerly competitive routes in South Wales and cut out expensive infrastructure such as on the Barry Railway's Barry-Porth route.

The cull of locomotive stock from the South Wales companies intensified. For instance, the Depression resulted in heavy withdrawals of the ex-Barry Railway mineral and shunting engines, many of the latter being sold for further service in industry. Engines which were not worn out, but merely redundant, the

GWR placed on a sales list. They were brought to Swindon Works and placed on sidings at its western end, locally known as the 'Dump'. Some engines spent months there, years even. Through the early 1930s the scrap lines at Swindon Works were filled by engines of standard GWR classes, such as the 'County' 4-4-0s and 4-4-2Ts. There was little sentiment attached to the engines of the companies absorbed by the GWR. Despite mostly being fitted with GWR-pattern taper boilers, the 0-6-0s and 4-4-0s of the Midland & South Western Junction Railway were extinguished during the mid and late 1930s. Similarly, the majority of ex-Cambrian Railways 4-4-0s and 0-6-0s had been taken out of service by the mid-1930s.

Whatever the effects of the Depression, it enabled the GWR to clear out all the non-standard types, to complete a process that had begun with the introduction of new standard classes in the 1920s. These had included the '56xx' 0-6-2Ts which had been built from 1924-28, and the '42xx' 2-8-0Ts multiplied in number between 1923 and 1930. Even these modern engines, which had been put into service in South Wales to replace engines from the absorbed railways, were affected by the loss in traffic experienced during the Depression. Twenty of the most recently built '42xx' 2-8-0Ts were stored from 1932 and orders for 20 others cancelled. The stored engines were rebuilt from 1934 as the '72xx' 2-8-2Ts for use on longer distance coal trains, and up until late 1939 further 2-8-0Ts were similarly treated.

Between 1930-34 the GWR went in for a remarkable revival of older engine designs. One of the Wolverhampton Works designs of 0-6-0 tank was adapted as the basis of a new series of 0-6-0 pannier tanks but with larger diameter wheels for use on auto (or push-pull) working. These were the '54xx' and '64xx' classes. Then the venerable '517' class 0-4-2T served to provide the general design of the '48xx' (later renumbered '14xx') engines, of which 75 were constructed from 1932 to take over existing or to introduce new auto-train services, at the time when the remaining GWR steam rail-motors were being withdrawn; the carriage portions of these vehicles were adapted to provide auto-trailers. Then the design of the '1361' class dock tanks, a small class dating from 1910, was updated with new Belpaire boiler and pannier tanks to result in the '1366' class of 1934.

Renewals continued of main line locomotive classes. The trend to replace the two-cylinder 'Saints' by four-cylinder 'Stars' was mentioned in Chapter 2 but through the 1930s both these excellent types of engine were progressively downgraded in their duties. The last 10 'Kings' had appeared in 1930 but two years later construction of 'Castles' resumed after a five-year break. There were just 40 'Castles' in 1927 but 115 *completely new* members of the class by the outbreak of World War 2. This qualification is needed for the GWR went in for rebuilding: five older 'Stars' became 'Castles' from 1924-29 and 10 of the most recently constructed 'Stars', Nos 4063-72, were reconstructed as 'Castles' from 1937-40. The engines were dealt with only when heavy repairs became required but just how much of the original fabric survived must be conjecture.

Some standard classes of engines were multiplied during the 1930s. The first of the mixed traffic 'Halls' had been produced in 1928 by rebuilding *Saint Martin* with smaller driving wheels, a change to the cylinder centre line and a new cab design. By 1930 there were 81 'Halls' but then, even during the depths of the Depression, production was stepped up, with another 80 of the class built from 1931-33. These saw off some of the GWR's 4-4-0s, such as the 'Atbara', 'City', 'County' and 'Flower' classes. The new engines were competent performers on express passenger and fitted freight work alike, and contributed to the trend to introduce longer through workings.

Below:
No 4999 *Gopsal Hall* (built at the height of the Depression, in March 1931) was photographed near Reading West with a Weymouth express. The first two vehicles are a GWR articulated twin dating from 1925/6. *Ian Allan Library*

Left:
'61xx' 2-6-2T No 6107 near Iver with a semi-fast train for Paddington. Next to the engine is a horsebox, and most of the train is made up of clerestory stock, including a former slip coach. *Ian Allan Library*

Left:
No 7810 *Draycott Manor*, built in December 1938, with the Swansea-Newcastle express, near Churchdown, between Gloucester and Cheltenham. The train was routed via Cheltenham, the Kingham avoiding line and Banbury to reach the Great Central main line. The train comprises LNER Gresley teak-bodied corridor stock including a restaurant car. *LGRP/Ian Allan Library*

Quantities of 2-6-2Ts also appeared during the 1930s, based on the '31xx' 2-6-2Ts which themselves had been modernised in the late 1920s as the '51xx'. The new engines, somewhat confusingly termed the '5101' class, were little changed in specification apart from some differences in their mechanical parts. One hundred of these engines were built to 1939, the majority finding employment on suburban passenger services in the Wolverhampton Division. Few of this series went to the London Division because, from 1931, the '61xx' 2-6-2Ts were built specifically for suburban services working into Paddington; these had boilers pressed to 272lb but few relatively minor differences as compared with the '5101' series.

There were other Large Prairie variants, too. Five of the (relatively) small-wheeled '3150' 2-6-2Ts were reconstructed as the '31xx' series in 1938/39, apparently for use on banking duties which had been one of their main tasks in original form, but instead the rebuilds gravitated to passenger work. Another example of reconstruction came with the '81xx' engines which reused parts of the main frames of the '51xx' engines but employed new front ends and boilers as well as new, smaller diameter driving wheels. Thanks to the outbreak of World War 2, just 10 of this intended 50-strong class were completed.

Had World War 2 not occurred when it did the GWR would have embarked upon some major reconstruction programmes. Many of the successful Churchward 2-6-0s were becoming due for major overhaul by the late 1930s and the Traffic Department had long felt that they would have been more useful had their boilers been of the larger No 1 type. There were 342 2-6-0s and the intention was to 'renew' them either as a mixed traffic 4-6-0 very similar to the 'Hall' class but with smaller 5ft 8in driving wheels, or as a lighter version with a new design of boiler.

The first of these two designs was known as the 'Grange' class and appeared from 1936. By 1939 there were 80 at work and although they were of almost entirely new construction they made use of the wheels and motion of the 2-6-0s. The 'Granges' were particularly popular for freight work, notably on fitted and partially fitted trains. The 'Manors' were the smaller of the two new 4-6-0 designs and the first took to the rails only in 1938. Just 20 were turned out in 1938/39 and seemed to spend most of their time on passenger workings. In postwar years their somewhat disappointing steaming qualities were successfully tackled and the 'Manors' were used on the Cambrian lines.

All the time that these main line engines were being built, the GWR pressed ahead with the construction of new 0-6-0PTs, largely to displace similar, older engines of Swindon and Wolverhampton design but also engines absorbed from the South Wales railways. The '57xx' 0-6-0PT was to become the mainstay of the fleet, and after 1945 there were no less than 863 in service. The first had been turned out by North British Loco in 1929 and more followed from other outside contractors during the most gloomy of the Depression years; their construction was financed under a Government scheme to relieve unemployment.

During the 1930s some classes such as the 'Aberdare' 2-6-0s and 'Bulldog' 4-4-0s may have suffered withdrawals but equally seemed everlasting. The former class were displaced on coal drags out of South Wales by the '72xx' 2-8-2Ts and by 1938 under half of the 'Aberdares' survived; the onset of war saw some returned to service instead of going for scrap. The 'Bulldogs' were replaced by '43xx' 2-6-0s which had lost work to newly constructed 'Halls'. Then, from 1936, the main frames of 29 'Bulldogs' were used with boilers from 'Duke' 4-4-0s to produce a replacement

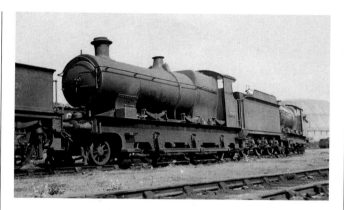

for the latter class. The largest number of the '32xx' (initially named after earls of the realm) went to Cambrian depots.

GWR coaching stock had changed greatly in design during the late 1930s and the design of some of the corridor stock resembled that of the LMS, with entry from end vestibules only, and large bodyside windows with sliding ventilators.

The GWR ended its prewar years on something of a downbeat note. All the Big Four companies saw a dip in traffic during 1938, following 1936/37 which had seen real improvement in their fortunes. The winter 1938/39 timetable saw the withdrawal of several express train workings on the GWR, including a number of cross-country trains, and two out of the three West of England-West Midlands expresses in each direction that were routed over the Honeybourne line. None the less, the late 1930s will be remembered for the high standards set by many of the cross-country expresses with which the GWR was involved, such as the Birkenhead-Bournemouth, Birkenhead-Margate/Hastings and Margate-Wolverhampton workings.

The SR in the 1930s

Investment in steam traction took a back seat on the Southern following the completion of the last 'King Arthurs' and 'Lord Nelsons' in the late 1920s. The emphasis was on other matters, principally electrification, the immense expansion of Southampton Docks during the early 1930s including construction of a graving dock and floating crane, the purchase of five new steamers for the Isle of Wight services, and acquisition of a proportion of the share capital of leading bus companies in Southern England.

Electrification reached Brighton on New Year's Day, 1933. By the end of that year, another couple of schemes had been approved by the Southern Railway Board, involving extension of electrification over 85 route-miles to Sevenoaks from Orpington and via Swanley, and along the South Coast from Brighton to Seaford, Eastbourne and Hastings.

There was a pause before further electrification schemes were prepared and submitted for Board approval. The first was the so-called Portsmouth No 1 scheme which would take electric trains to Portsmouth via Woking, Guildford and the Portsmouth Direct line, to Farnham, and from Staines via Virginia Water to Weybridge. This project was being

Above:
An aerial view of the Southern Railway's Southampton Docks, dated July 1929, before the completion of major improvements at the port. Centre of the picture and berthed at the Ocean Dock are two major ocean liners of the time, the four-funnelled *Olympic*, beside which is the *Leviathan*, and further away is the *Mauretania*. Above and top right of the *Mauretania*, the New Docks are taking shape. *Ian Allan Library*

considered by the SR's directors when the Government announced financial facilities in support of new works undertaken by the main line railways. As approved at the end of 1935, the Portsmouth No 1 scheme now included the extension of electrification from Farnham to Alton.

The approval of electrification schemes on the Southern was not a foregone conclusion. More than one scheme was withheld for further deliberation because the projected return on capital was not promising enough. The often substantial increase in train miles operated by electric multiple-units depended on attracting many more passengers, while the schemes were necessarily required to be completed within authorised financial limits. The Southern's management met both these challenges and as a result of their careful planning and efficiency, railway electrification was made to look both simple and effective.

July 1936 saw the Portsmouth No 2, or Mid-Sussex, electrification authorised; then, in October the same year, the Reading line scheme was approved. The extension of electrification to Gillingham (Kent) and to Maidstone was authorised in October 1937. This summarises very briefly the final prewar SR electrification projects, the last two mentioned being completed in July 1939. Other schemes included the construction of a new South Western suburban line, to Chessington South. One scheme that did not pass the test, the proposed electrification from Sevenoaks via Tunbridge Wells to Hastings, failed to gain Board approval in late 1937.

After the last 'King Arthurs' and 'Nelsons' in 1927/28 next came CME Richard Maunsell's remarkable 'Schools' class 4-4-0s from 1930, well-timed indeed, for these engines held the fort until the commencement of electrified services on the Portsmouth Direct line in 1937. They remained at work until the late 1950s on the London-Tunbridge Wells-Hastings route, the proposed electrification scheme for which had not passed muster.

Back in 1928, the SR Board had approved the construction of 25 passenger engines whose type was not however specified; of these, 10 were 'Schools'. A further 20 of this class were approved in 1931 but, because of the spinning-out of the building programme during the Depression years, the last of the 20 was not completed until 1934. Another 20 'Schools' were authorised in 1932 but the order was later reduced to ten, the last of these being accepted for traffic in mid-1935.

This gives some clue to the Southern's tight control on new steam locomotive building. The 'Z' class eight-coupled heavy shunting tank engines appeared in 1929. Eight were built but 10 more were dropped from the 1930 building programme. A variety of other designs proposed by Maunsell failed to materialise in the 1930s for reasons either of cost or because the civil engineer was not prepared to accept them.

Left:
Lewes station, c1935, showing the recently built sub-station in the mid-background, together with a new electric multiple-unit for the Eastbourne/Hastings electrification approaching on the left of the picture. An ex-LBSCR 0-4-2T is waiting with a motor-train in a bay platform on the right of the picture. *Ian Allan Library*

Left:
The Mid-Sussex line (or No 2 Portsmouth) electrification was inaugurated in July 1938. On 25 June that year, the 3.20pm Victoria-Bognor Regis Pullman car express passes near Hackbridge, between Mitcham and Sutton, behind ex-LBSCR 'H2' 4-4-2 No 2425 *Trevose Head*. *E. R. Wethersett*

Above:
Malden Manor station is on the Tolworth-Chessington South branch which opened as an electrified line on 29 May 1939. The concrete platform canopies incorporated cold cathode lighting. *Ian Allan Library*

Opposite:
The Southern Railway's 'Bournemouth Limited' express ran nonstop between Waterloo and Bournemouth Central in 2hr exactly, and was introduced in July 1929. Here it is with 'King Arthur' 4-6-0 No 789 *Sir Guy*, at the head of a train which includes 'Ironclad' corridor stock built by the LSWR in 1921, a design initially adopted by the SR after Grouping. *Ian Allan Library*

The more notable designs comprised a four-cylinder Pacific, a mixed traffic 2-6-2, a Beyer Garratt, a heavy goods 4-8-0, and a two-cylinder 2-6-2T. The failure of the last of these proposals meant that the Southern was left with a number of elderly passenger tank engines which were destined not to be replaced until the early 1950s. Indeed, not a single new type of steam engine was introduced by the SR between 1932 and 1935. The need for heavy goods engines was met eventually by building 10 'S15' 4-6-0s in 1936.

A policy of 'make do and mend' applied equally to the Southern's suburban electric rolling stock which was largely produced by rebuilding steam-hauled coaches. Nowhere was 'make do and mend' more true than on the Isle of Wight. A. B. MacLeod took charge of the Island's Locomotive, Carriage and Wagon Department in 1928 and this enthusiastic railwayman was subsequently also made responsible for the traffic and commercial departments.

This period saw a number of improvements to the Island's railways, including construction of a new locomotive running shed, modernisation of the repair depot, and a sprucing-up of the locomotive stock. More engines came over during the early 1930s to increase the Island's stock, including two more ex-LSWR 'O2' class 0-4-4T, a type first transferred to the Island in 1923, four ex-LBSCR 'E1' 0-6-0Ts, and a number of secondhand bogie coaches. More 'O2s' and more bogie coaches were transferred from the mainland during 1936 until there were 27 locomotives and 85 bogie coaches to meet a continuing increase in traffic. By now, the locomotives of the old Isle of Wight railway companies had been withdrawn, and some of the 'Terriers' returned to the mainland, the strengthening of bridges having allowed the larger 'O2s' to work most of the Island's trains. The last of the former IoW company engines was Beyer Peacock 2-4-0T No W16 *Wroxall* which in its 61 years' service had run nearly 1.5 million miles - not bad for an engine on an island railway. Similar W13 *Ryde* had been sent to Eastleigh in 1934 and at one time was expected to be preserved, only to be lost to an early wartime scrap drive.

Electrification combined with the limited acquisition of new engines during the 1920s allowed the SR to withdraw older classes of engines during the early 1930s. These included, from former South Eastern or SE&CR stock, the last domeless 'F' (1930) and 'B' 4-4-0 classes (1931) and the last domeless 'O' 0-6-0 (1932) while, of the LBSCR classes, the last Stroudley 'B1' or

Above:
Ryde St Johns Road station, after modernisation during the 1920s by the SR and resignalled using a signalbox that had been moved from Waterloo Junction, London. Note the newly built platforms with concrete sections and fencing panels produced by the SR's own concrete works. *Ian Allan Library*

Below:
Trains for the Kent Coast passing Bromley South in July 1937. Transferred from the Western Section, ex-LSWR 'T9' 4-4-0 No 307 is about to be passed by 'King Arthur' 4-6-0 No 763 *Sir Bors de Ganis*. Fitted specially with a six-wheeled tender, No 307 remained on the Eastern Section from 1925-40. *H. C. Casserley*

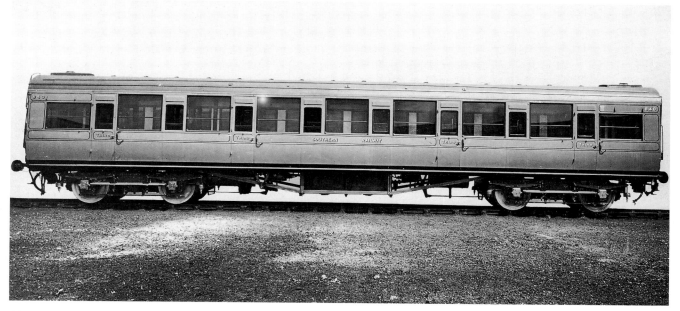

Above:
SR corridor third No 840, built in 1929, showing the corridor side which, in common with all SR corridor stock post-1929, sensibly had windows extended to the cant-rail to make it easier for passengers to see out. The vehicle has been specially prepared for this works photograph with lined-out works grey paintwork; the standard SR sage-green livery was somewhat sombre.
Ian Allan Library

Below:
'Lord Nelson' 4-6-0 No 862 *Lord Collingwood* approaches Dover Marine with a boat train from Victoria, c1937. The engine had been modified with Kylchap blastpipe and double chimney in 1934, one of various attempts to achieve consistent performance from the 'Lord Nelsons'. When Bulleid became CME during 1937, the cylinders and draughting of these engines were improved.
B. K. Cooper

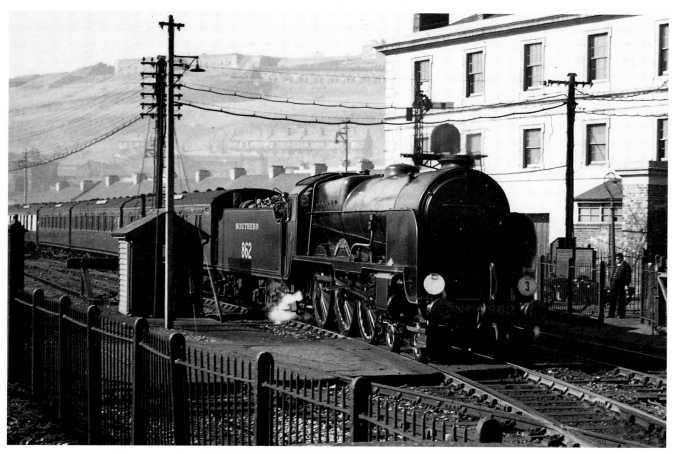

'Gladstone' 0-4-2 went in 1933, along with the last rebuilt Billinton 'B2x' 4-4-0. Of the ex-LSWR classes, the only significant example to be eliminated was the Drummond Class 8 4-4-0 whose final member was scrapped in 1938. Although reduced in number, the last surviving Adams 4-4-0s eked out their existence until the war years. The 'C8s' apart, the various classes of Drummond 4-4-0s remained hard at work through the 1930s, some of the 'L12s' and 'T9s' 4-4-0s having been redeployed to other SR Sections during the 1920s and 1930s.

The Drummond 'M7' 0-4-4Ts increasingly lost their duties on the Western Section's London suburban services to electric multiple-units, and from 1930 a number were fitted with the LBSCR's system of compressed air control for working push-pull trains. It was only after World War 2 that they strayed far from former South Western metals.

As explained in the previous chapter, no new non-corridor steam stock was built by the Southern Railway but the construction of Maunsell's corridor stock continued through the 1930s. These must be reckoned to be among the most comfortable of any British steam age coaches. There were two main versions of this corridor stock, one with slab bodysides and lacking guard's duckets that suited the restrictive loading gauge of some Eastern Section lines, and the other with rounded 'tumblehome' bodysides for use on the Western and Central Sections. The most severe restrictions were imposed by tunnels on the Tonbridge-Hastings line, and the body width of this so-called Hastings stock was no more than 8ft 0¾in. This type excepted, a number of open layout coaches were constructed, suitable either for dining use when they were marshalled next to restaurant cars, or for party excursions. From 1934-5, the exterior style of Maunsell coaches changed: at first to coaches somewhat fearsomely adorned with rows of screwheads on their body panelling, and in appearance unlike any of the SR's electric stock, but then to designs of open and side-corridor coaches which closely resembled vehicles built for the 4-COR electric multiple-units.

The last of these coaches went into traffic after Oliver Bulleid had taken over as Chief Mechanical Engineer of the SR in October 1937. Much of the London suburban network had been electrified, as well as the Brighton and Portsmouth main lines, but there remained many important steam-worked main line services, to the Channel Ports and Kent Coast, to Bournemouth and Weymouth, and to Salisbury and further into the West of England. Premier trains of the early 1930s included the 2hr 'Bournemouth Limited' that had been introduced in 1929 and, two years later, the all-Pullman 'Bournemouth Belle'. Continuing

electrification threatened to show up the deficiencies of the steam stock, made all the worse by the SR Chief Civil Engineer's refusal during the first half of the 1930s to accept Maunsell's designs for large express locomotives.

Bulleid was an all-round locomotive and rolling stock engineer and was possessed with a gift for finding original and unconventional solutions which seems to have influenced his appointment as CME of the SR. Having had a close look at matters Southern, Bulleid obtained authority in March 1938 to build 10 new express passenger engines, one of their main intended duties being the Channel Ports' boat trains. By then, Eastleigh drawing office had prepared an outline design for a new streamlined Pacific, but it risked being unacceptably heavy, so thoughts then turned to a 2-8-2, only for the design of a Pacific to be put in hand the following year. The order for the first 10 of this type was placed with Eastleigh and its suppliers in mid-July 1939. The first of these, the 'Merchant Navy' class, appeared in 1941.

Bulleid meanwhile looked hard at the existing Southern express engines, and decided to improve the always somewhat unpredictable performance of the 'Lord Nelsons'. The Lemaître-pattern of multiple-jet exhaust and wide chimney was applied to the 'Nelsons' which also received new cylinders, both modifications being effected from 1939 onwards. Just under half of the 'Schools' class were also fitted with Lemaître exhausts and wide chimneys which made them even more effective and fast-running engines. One of Bulleid's other innovations had been a vivid shade of malachite green which made its appearance as the SR steam locomotive livery from 1938.

The only new steam engines put into service by the SR at the end of the 1930s were the modest and simple 'Q' class 0-6-0s, of which 20 were built in 1938/9 to a design dating from Maunsell's time. In the early months of the war, Bulleid experimented with one of the 'Qs' fitted with wide chimney and Lemaître multiple-jet blastpipe. No Bulleid-influenced rolling stock entered service before World War 2 but construction of sets of steam-hauled corridor coaches had been put in hand in 1939, their completion being deferred until 1945-6.

The LMS in the 1930s

In 1930 the LMS was preoccupied with the challenge of reducing its expenditure to meet the reduction in revenue caused by the Depression. By 1930, what was there to crow about on the LMS? From 1927, the first 50 'Royal Scot' 4-6-0s had entered service and, among other developments, had made possible nonstop running with the 'Royal Scot' express over the 299 miles between Euston and Carlisle. Another 20 'Royal

Above:
A 1933 picture of a Manchester-St Pancras restaurant car express, seen near Chee Tor, Miller's Dale in the Peak District. The engine is ex-Midland Railway Compound 4-4-0 No 1028 which is at the head of LMS corridor stock. *Ian Allan Library*

Scots' were built in 1930. Thirty more Beyer Garratts were constructed for the Midland lines' coal traffic in the same year, and 1930 also brought the first of a new 2-6-2T class which was never to be renowned for its performance.

As the 1930s progressed, the LMS attempted to pull together its widely spread empire. It concentrated on its long-distance passenger and freight services, particularly on its Western Section where there were 51 nonstop runs of over 100 miles in 1930, as compared with 41 in 1914; noticeably, the totals for its other Sections in 1930 compared unfavourably with 1914. Yet, none of the long-distance LMS runs was particularly fast.

Economy was the guiding principle for much of its network. At least twice in its lifetime, the LMS contemplated closing the lines north of Inverness, and even the entire former Highland system. In a 1936 report, the LMS admitted that the Highland section was barely making a profit although calculations showed that the Company risked losing much valuable traffic if closure was effected. The conclusion was reached that on balance the Highland section should be retained.

The Midland Railway had not managed to electrify the London, Tilbury & Southend Section as the Company had promised in 1912 and these lines saw little in the way of modernisation until the first half of the 1930s, apart from receiving a few new sets of coaches and a modernised version of the LT&SR 4-4-2T. Changes came with the arrival of new three-cylindered, taper-boilered 2-6-4Ts and more new coaches, the construction of additional and electrified slow lines between Barking and Upminster in 1932, accompanied by the LNER's remodelling and resignalling of the Fenchurch Street terminus. This station was used by the LMS Southend and Tilbury trains and its improvement led to an enhanced train service.

There was no doubt that the 50 'Royal Scots' had been ordered in a hurry from the North British Loco company at the end of 1926. The builder initially promised to deliver the entire order by the end of the summer of 1927 but did well to provide all 50 engines by December 1927. Probably the design and production of the 'Royal Scots' was too rushed. From the mid-1930s major changes were needed to the running gear of the 'Scots' including new axleboxes, and the modifications were duly applied.

Despite their defects the 'Scots' served the LMS well, particularly with the haulage of heavy Western Section expresses. Many of these trains loaded to well over 400 tons which the big 4-6-0s mostly worked unassisted. Some of the West Coast expresses featured a complicated marshalling of coaching stock at stages on their journeys, none more so than the 6.45am Aberdeen-Euston during its 540-mile journey; its

Right:
The Big Four sought economies in operation during the 1930s, such as by abolishing signalboxes in favour of intermediate block colour-light signals, typified by this LMS example at Weedon, Northamptonshire on the West Coast main line, and photographed when new in 1935. *Ian Allan Library*

Below:
Another development of the late 1930s was the use of road/rail vehicles, such as these LMS edible oil tankers loaded on to a flat wagon. *Ian Allan Library*

schedule provided for numerous connections to be made out of and into the train at principal stations. In later days, the 'Royal Scot' working through from Carlisle with the 6.45am was faced with a just faster than 60mph start-to-stop timing from Blisworth to Euston on the last leg of its 299-mile run.

In the late 1930s, trains such as the 11.50am Euston-Manchester were booked to load to 15 coaches for 481 tons on certain days of the week, and the 6.5pm Euston-Liverpool to 443 tons. Some West Coast expresses conveyed loaded six-wheeled milk tank wagons. One example was the 10.5am Aberdeen-Euston: marshalled next to the engine was a milk tanker which worked throughout from Aberdeen to London while between Preston and Nuneaton the express conveyed another milk tank, in this case consigned from Garstang to Cricklewood, by way of Nuneaton and Leicester.

By the autumn of 1935, 12 Stanier 'Princess Royal' Pacifics were at work on principal West Coast express trains. The remit to the CME had been to produce an engine capable of working from Euston to Glasgow with 500-ton trains, unassisted over Shap and Beattock banks. The first two of the class, Nos 6200/01, had been built in 1933, and no more were ordered for the next couple of years. As a result of experience with Nos 6200/01 a number of improvements were incorporated in the later members of the class, particularly when it came to the design of their boilers, the later examples of which had an increase in superheater elements and other changes. A more adequate tender was also provided. There was one other Pacific, No 6202, the unique turbine-driven Turbomotive of 1935, which must be considered one of the most successful genuinely experimental steam locomotives and had lengthy spells working the heaviest Euston-Liverpool expresses.

Before the introduction of the Pacifics, the LMS had needed more than 70 'Royal Scots' to cover all its principal express turns on the Western Division. From 1926, a number of attempts had been made to get the best from the 'Claughtons', including fitting some with Caprotti valve gear, by use of an enlarged boiler (both modifications being incorporated into some members of the class), changes of valve rings, and the latest developments in draughting. On test many of the class performed in exemplary fashion but defects in the detailed mechanical design of the 'Claughtons' let them down and their reliability was poor. Eventually,

Above:
The LMS worked hard to make the best of the 'Claughtons' whose reliability was disappointing: in the last six months of 1930 there were 60 cases of hot trailing axleboxes with members of the class. Ironically, this commercial photograph shows a wheeldrop in a Midland Division shed in the early 1930s with a locomotive wheelset raised to rail level for repairs, from 'Claughton' No 5973 which can be seen in the background. *Ian Allan Library*

it was decided to rebuild them as virtually new three-cylinder engines. Once the first two had been outshopped, the strategy changed to building new engines, although perhaps the accountants were pacified by continuing to classify the engines as 'rebuilt' instead of 'new'. Forty of what from 1936 became known as the 'Patriot' class were nominally, but only nominally, rebuilds of 'Claughtons' and, once Stanier's new Class '5' and '5X' 4-6-0s began to appear in quantity after 1934, the 'Claughtons' in original and fettled-up condition, with old-style or large-diameter boilers, were withdrawn. By 1937, just four remained out of the original total of 130. The 'Patriots' eventually totalled 52 engines, 10 being regarded as entirely new, and the balance of five on the official sanction being completed instead as the taper-boilered '5Xs' (later the 'Jubilee' class).

Meanwhile, a huge programme of locomotive construction was proceeding on the LMS, following the issue of tenders in November 1933 for the supply of two- and three-cylinder 4-6-0s. At first, 50 of the three-cylinder engines were ordered from North British Loco and 50 two-cylinder 4-6-0s from Vulcan Foundry.

These were the first '5X' engines (Nos 5557-5606) and '5s' (Nos 5020-69) respectively. On receipt of bids against the next set of tenders issued in July 1934, the LMS considered that the outside builders' prices were too high and some of both 4-6-0 classes were constructed instead by the Company's workshops.

With the next round of tendering, the LMS ordered 100 engines from Armstrong Whitworth (Class '5s' Nos 5125-5224), and 50 from Vulcan Foundry ('5s' Nos 5075-5124). In November 1935, the company went out to tender for a grand total of 369 Stanier engines - 227 Class '5s', 69 2-8-0s and 73 2-6-4Ts, their cost of construction being part-financed by one of the

Government assistance schemes of the time. The LMS drove a very hard bargain with Armstrong Whitworth when it placed an order for the 227 Class '5s', Nos 5225-5451.

When appointed CME of the LMS in 1932, Stanier had brought with him a number of locomotive design features from the GWR and Swindon Works. Not all worked well in LMS conditions. Some general features applied to both the two-cylinder '5s' and three-cylinder '5Xs' were changed to some degree in ensuing years. Having said that, the '5s' entered service and generally performed well but the '5Xs' did not steam satisfactorily. Before 1934 was out, members of this class were being subjected to dynamometer car tests alongside the 'Patriots'; the comparison was not flattering as the older engines were shown to be more economical. These results influenced the construction of further '5s' and '5Xs' incorporating changes in the design of the boilers, including a greater degree of superheating. Testing of both '5X' and '5' engines continued throughout 1935, such that the boilers for both classes were further redesigned. It was not until 1937, however, that a 'Jubilee' incorporating the progressive improvements made with boiler design turned in a standard of performance expected of engines in its power group.

The '5s' - which become known as the 'Black Fives' to distinguish them from the red-painted 'Jubilees' - proved very capable over the full range of duties and soon were to be found on all parts of the LMS, including its Northern Division (Scotland). The only impediment to their widespread use on the network was the presence of weak bridges which in the coming years were strengthened on lines as far apart as the Callander & Oban line and the Somerset & Dorset.

Whatever the limitations of the 'Jubilees' in their original form, when the accelerated and greatly revised express services on the Midland Division were introduced in October 1937 the 'Jubilees' and '5s' were put to the test and proved fully capable of keeping to the mile a minute timings. 'Jubilees' also did well on the 3hr Glasgow-Aberdeen expresses introduced in the summer of 1937, as well as on the accelerated cross-country services on the Bristol-Sheffield route which were a feature of the September 1937 timetable.

Other Stanier classes multiplied in number in the late 1930s were the 2-8-0s and 2-6-4Ts. The 2-8-0s were straightforward engines which soon became favourites with the enginemen and operating staff alike - not perhaps surprisingly as they easily outclassed the Midland/LMS-built '4Fs' and none too brilliant Fowler 0-8-0s - but only 125 had been put into service before the outbreak of World War 2. Some of these engines were sent overseas on military service and a proportion never returned. Of the

Above:
Stanier 'Black Fives' were increasingly passed to work over sections of the LMS network. In 1938, they were cleared to work over the Somerset & Dorset line, and here is No 5432 working a stopping train near Binegar, in the June of that year. *Ian Allan Library*

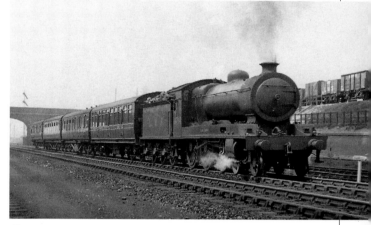

Above:
From 1934, the Highland 'Clan' 4-6-0s began working over the Callander & Oban line, but just before World War 2 they were replaced by Stanier '5s'. In 1938, 'Clan' No 14764 *Clan Munro* passes Balornock shed, Glasgow with an Oban-Glasgow Buchanan Street working. *H. C. Casserley*

Above:
The Stanier 2-6-4Ts were a vast improvement on most of the secondary passenger engines they replaced. No 2458 is seen working a Buxton-Manchester train formed of modern LMS steel-panelled non-corridor stock, and passing Fairfield Golf Links Platform, near Buxton, in 1939. *Ian Allan Library*

2-6-4Ts, Nos 2425-94 and 2537-2652 had been built before production stopped in advance of the war. Whereas the 2-6-4Ts built for the Tilbury section in 1934 had three cylinders, the others were two-cylinder and proved to be one of the best all-round tank engines in Britain. They were unusual in being fitted with water pick-up gear for use in both directions of running so that they could take water from track troughs instead of increasing the time spent at station stops. The Stanier Class '4' 2-6-4Ts established a good reputation on the tightly-timed suburban trains out of London to Watford and beyond. Less successful were the Stanier '3' 2-6-2Ts which were under-boilered for the jobs they were supposed to cover.

By the late 1930s the LMS scene had changed, at least on the heavily trafficked main lines, thanks to the rapid construction of the Stanier standard locomotive classes, and the equally rapid building rates of Stanier period side-corridor and open steel-panelled coaches and non-corridor vehicles. In many ways it was difficult to fault the Stanier period coaches which rode well, despite their customary and somewhat short 57ft length. Yet the overall scene of the late 1930s' LMS main lines was a trifle drab, not helped by the fact that engines and coaching stock were often indifferently cleaned, which was partly a consequence of their intensive use.

Punctuality was perhaps a different matter: the LMS operating department was very enthusiastic when it came to clipping minutes off schedules and instituted an 'On Time' campaign from the mid-1930s which was reinforced by a staff newspaper of the same name. Chief Operating Manager, C. R. Byrom, issued a clarion call to train and station staff: 'Every Train To Time On Every Day'. The *On Time* newspaper featured league tables of the different LMS Divisions' achievements in passenger train timekeeping and highlighted instances when engine crews had worked hard to recover lost time. The company continued to accelerate its express passenger trains, and by the summer of 1938 it was way ahead of the other Big Four companies. The LMS boasted 97 runs each weekday which were timed at over 58mph, with a total mileage of 9,000.

Some aspects of the pre-Grouping companies' characters remained unchanged until the LMS was engulfed by war, and even beyond. Perhaps the most magnificent emblem was the former London & North Western Railway royal train, some of whose vehicles dated from 1902. After Grouping it is said that the LMS wanted to repaint the royal train vehicles in overall crimson lake but that King George V

Below;
An impressive line-up at Camden motive power depot in London with, left to right, 'Jubilee' 4-6-0 No 5559 *British Columbia*, No 6100 *Royal Scot*, streamlined 'Princess Coronation' No 6224 *Princess Alexandra*, unstreamlined No 6230 *Duchess of Buccleuch*, and another streamlined Pacific.
Science & Society Picture Library LMS 8633

The LMS in Northern Ireland: NCC 'U2' 4-4-0 No 75 *Antrim Castle* (built by North British Loco, 1924) working a Larne-Belfast York Road train formed of steel-panelled non-corridor stock. *LPC/ Ian Allan Library*

discouraged the idea. The train retained its LNWR livery of bluish white upper panels and dark carmine lake lower sections even though some of the vehicles had been modernised in the 1930s by receiving new underframes. As World War 2 proceeded, it was thought that the royal train was altogether too visible, so King George VI consented to its vehicles being repainted in standard LMS crimson lake.

On at least one occasion, one of Stanier's streamlined 'Princess Coronation' Pacifics painted in crimson and gold striped livery appeared at the head of the venerable royal train and surely provided bystanders with an unforgettable image. The later Pacifics had appeared from 1937 with the introduction of the 'Coronation Scot' of the LMS which in each direction was booked to cover the 401 miles between Euston and Glasgow Central in 6½hr. While the initial batch of Pacifics, Nos 6220-24, were shrouded in a streamlined casing, the nine coaches in each of the two train sets used were of unstreamlined standard stock. Coaches and engine alike were finished in a striking livery of blue with parallel white stripes.

A casual observer of the time would have noted this modern, eye-catching train passing some relatively unchanged ex-LNWR and Caledonian stations - none too recently painted at that! - and a plethora of North Western signals and signalboxes.

Northern Ireland's Railways in the 1930s

If the Big Four railways had a hard time competing with road transport, it was doubly worse in Northern Ireland. Road passenger services may well have been regulated by a system of licences but the carriage of goods was completely unregulated. No less than 700 omnibuses were in operation by 1933, owned by 50 or so operators. When it came to rural and local services the Province's railways were hard-put to be competitive, but they could exploit their advantages on longer distance main line services.

With its absorption of the Midland Railway the LMS also acquired a railway system in Northern Ireland. This was the Northern Counties Committee (NCC), comprising 279 route-miles of 5ft 3in gauge track and 78 miles of 3ft gauge. The NCC was an amalgamation of various Irish lines, principally the Belfast & Northern Counties Railway which had amalgamated with the Midland Railway in 1903. The B&NCR had progressively swallowed up a number of other railways in Northern Ireland between 1855 and 1924.

Above:
The Great Northern Railway of Ireland: compound 4-4-0 No 84 *Falcon* working a Belfast Great Victoria Street-Dublin Amens Street express. *LPC/Ian Allan Library*

After the formation of the LMS, Derby's influence became ever more apparent in the design of locomotives and coaches supplied to the NCC. There were three 'V' class 0-6-0s built at Derby in 1923 which had a definite Midland appearance, followed from 1924 by the 'U1' and 'U2' 4-4-0s, again externally very similar to the Midland's Fowler Class '2' 4-4-0s but with 6ft driving wheels. The builders of the earlier engines of the class were North British Loco but later examples were constructed at Belfast York Road Works. From 1933, the 'W' class 2-6-0s were introduced, based on the general design of the Fowler 2-6-4T, but having 6ft driving wheels. Fifteen of these engines, the majority named, entered traffic up to 1942. The earliest of the class were supplied complete from Derby Works but later engines came as kits of parts, to be erected at York Road.

The LMS made a number of improvements to the NCC. Greenisland Viaduct opened in January 1934 and, by removing the need to reverse at Greenisland, it permitted trains for Coleraine and Londonderry to be accelerated between the NCC's York Road terminus at Belfast and Ballymena. Booked to cover the 31 miles from Belfast to Ballymena in as many minutes, the 'North Atlantic Express', which ran on Saturdays between Belfast and Portrush, was claimed to be the fastest train in Ireland. It ran the 62½ miles to Portstewart nonstop in 69min.

The Great Northern Railway of Ireland (GNR [I]) had a larger network of lines than the NCC and after 1922 straddled the border between Northern Ireland

and the Irish Free State. Its main line ran from Dublin, via Dundalk and Portadown to Belfast Great Victoria Street station. With the rebuilding of the Boyne Viaduct, Drogheda, in 1932, the express trains were accelerated and, including three intermediate stops, bookings of 2hr 20min between Belfast and Dublin now appeared in the timetable.

To work these expresses, the GNR (I) Locomotive Superintendent, G. T. Glover, had introduced five powerful three-cylinder Compound 4-4-0s. They were numbered 83-87, and carried the names *Eagle*, *Falcon*, *Merlin*, *Peregrine* and *Kestrel*. Even when the Dublin-Belfast timings had to be eased in 1934, the GNR (I) still offered the fastest running in Ireland. The GNR's locomotive fleet numbered 201 in 1930, the most numerous types being 0-6-0s (98), 4-4-0s (63) and 20 4-4-2Ts, the last principally for working the Belfast suburban trains.

After World War 1, one-third of the locomotive stock of the Belfast & County Down Railway (BCDR) was scrapped. Four new 4-4-2Ts were acquired in 1921, and four 4-6-4Ts were acquired in 1920/1. Impressive though they might have been in appearance, the 4-6-4Ts were not good performers and in any case were restricted to the Bangor line. Suburban traffic was the lifeblood of the BCDR and remained important during the interwar years.

Ireland's railway strike in 1933 contributed to the first real pruning of the NCC's and GNR (I)'s systems. The railways' competitive position was further weakened by the formation in 1935 of the Northern Ireland Road Transport Board. This promised to co-ordinate road and rail transport but its actions fuelled many people's convictions that it was an attempt to favour road at the expense of the railways.

The LNER in the 1930s

The LNER's Ordinary General Meeting held at the Hotel Great Central on 7 March 1930 began with good news. Holders of the Company's ordinary stock would receive a dividend. The rest of the tidings in the Report of the Directors were mixed. Salaries and wages had just been reduced by 2½%, the Company had purchased a financial interest in a number of bus companies, the Government's remission of passenger duty meant that a number of investment schemes could go ahead, the first of three fine new steamers

for the Harwich-Hook of Holland route had been put into service and the others would follow imminently, and the LNER's Bill for new dock works at Grimsby had received the Royal Assent.

This was, perhaps, a typical enough set of developments for one of the Big Four companies but now that the Depression had taken hold the LNER's financial position was even more under strain than usual. More wage reductions, loss of jobs, cut-backs and deferments would follow.

During 1930, work was in progress on the electrification at 1,500V dc of the Manchester, South Junction & Altrincham Railway which was a joint enterprise with the LMS. Electrified working began in May 1931. Maybe this was a sign of things to come. It was not until the tail-end of the 1930s that the LNER began preliminary work on the 1,500V dc electrification of its Manchester, Sheffield and Wath lines, and the Liverpool Street-Shenfield suburban electrification. In the meantime, the LNER remained very much a steam railway. Early in 1930 it had had to

Right:
In the aftermath of the General Strike, the LNER purchased more Robinson 2-8-0s from the Government and used them to replace less efficient engines, such as the ex-Hull & Barnsley Railway 0-8-0s. The arrival of the Depression and the loss of traffic made more freight engines redundant. This 1934 view of the former H&B repair shop at Hull Springhead depot shows an 'O4' on the far left, an ex-NER 'J27' 0-6-0, ex-NER 'Q6' 0-8-0, and an ex-H&B 'N13' 0-6-2T. *T. Rounthwaite*

Below:
A trick photograph - *three* new steamers for the LNER's North Sea services from Harwich Parkeston Quay? No, a montage employing a single image of the *Amsterdam* which entered service in 1930! *Ian Allan Library*

concede that 'it may be some time' before the electrification of the Great Northern and Great Eastern suburban services could proceed. Instead, new loco-hauled steam stock was ordered.

Attempts were made to consider an alternative to the conventional steam-hauled train. Gresley produced a design for an early diesel locomotive, then serious consideration was given in 1932 to use of Sentinel-Cammell steam railcars on the King's Cross-Hitchin-Cambridge service. Revenue had fallen as a result of increased road competition, mainly from coaches. Just how the steam railcars would have fared is open to doubt. At any rate, the LNER opted for lightweight, steam-hauled trains that featured buffet cars. These, the famed 'Beer Trains', proved a huge success. Buffet cars were just one of the innovations adopted by the LNER - and of course the other railways - to make train services more appealing. Buffet cars were used when the provision of a full restaurant car was uneconomic.

The demand for high-class services may have taken a bit of a knock at the onset of the Depression but soon recovered. In 1930, the LNER introduced what it termed 'Super First' coaches on its East Coast expresses. It did so all the while keeping a beady eye on the LMS which also put some splendidly plush first-class coaches on the 'Royal Scot' that year. The LNER stole a march on the LMS by providing pressure ventilation and heating in its 'Super Firsts' whose pastel paint finishes compared favourably with the somewhat overbearing varnished wood panelling in the interior of the LMS vehicles.

Other attempts to earn revenue were aimed at ordinary folk, such as the first 10 camping coaches introduced by the LNER during 1933. Placed in sidings at branch line stations they could be hired for 50s a week. Also in 1933, it was decided to build five new 12-coach trains 'for excursion purposes', these being made up of new plywood-panelled coaches including a buffet car. The excursion trains were a way of attracting new business to rail, and ensuring that it

Below:
On 23 July 1934, a Cambridge-King's Cross 'Beer Train' accelerates past Meldreth behind ex-GNR 'C1' Atlantic No 4436. There are no more than five coaches in the train including the buffet car. *E. R. Wethersett*

remained superior to the road coaches for trips to the seaside and major sports events.

These examples of special services, whether for first class passengers or excursionists, serve to demonstrate that the Big Four railways did their best when trying to reverse the decline in revenue during the Depression. Whatever the developments in mechanical engineering, and in particular new locomotives, the real challenge was to pull in business.

On the freight side, the same approach was required. *How the LNER 'Expresses' Freight*, a 1932 sales brochure for freight forwarders, features selected express freight trains and advises of the latest time for acceptance of traffic at the starting station, and the times for collection at destination. Traders making use of the 'Three-Forty Scotsman', the crack London to Scotland service leaving King's Cross at 3.40pm, could bring their traffic to King's Cross as late as 2pm the same day, and expect it to be ready for delivery in Edinburgh at 7am the next morning, and in Aberdeen at 11.30am. A later LNER facility was Green Arrow by which, on payment of a registration fee, a consignment of full-load traffic would be monitored throughout its transit.

The LNER aimed to improve the handling of freight traffic by building new marshalling yards. The mechanised down yard at Whitemoor was completed in 1933, the up yard having been dealt with in the late 1920s. The new yard at Mottram, 11 miles from Manchester, was opened in 1935 to expedite the sorting of wagons bound for that city and beyond, and the same year a new mechanised yard at Hessle, near Hull, began operations.

Above:
LNER plywood-panelled Tourist stock in use - an up excursion train pauses at Nottingham Victoria station, c1935/6. The engine is 'K3' 2-6-0 No 1164. *T. G. Hepburn/Rail Archive Stephenson*

Below:
An LNER Scammell mechanical horse negotiates a tight bend at Farringdon Street Goods Depot, London, in 1934.
Ian Allan Library

Some of the most attractive locomotive designs to be inherited by the LNER were the Robinson and Pollitt classes of the Great Central Railway. Of the 1,358 ex-GCR engines, there was hardly an ugly or badly proportioned one among them. Not long after the Grouping had taken place, enthusiasts were horrified when ex-GCR engines began appearing with disfiguring boiler mountings and their cabs cut down. The reason was simple: Robinson's engines made full use of the GCR loading gauge and had to be altered to conform to the more restricted LNER Composite Load Gauge. The built-up chimneys took no account of the flowing lines of the engines and to enthusiasts they were 'flowerpot' chimneys.

There was another reason for replacing the elegantly fashioned GCR chimneys, for they were prone to cracking. Whether the LNER responded to enthusiasts' criticisms is unknown, but from the mid-1930s ex-GCR engines began to be fitted with new designs of chimney and dome.

The LNER naturally introduced its own practices, and as a result made modifications to engines inherited from its constituent companies. There was a fashion to fit feed-water heaters to some classes. The idea was to preheat water passing to the boiler by use of the exhaust steam from the cylinders which was mixed with cold water from the tender tank. The apparatus mounted on the engine's boiler comprised pumps, and mixing and settling chambers. Heated water was forced into the boiler by one of the pumps. Why bother with this extra equipment? The manufacturers claimed that their feed-water heaters achieved savings in coal and water, but that was similarly true of exhaust steam injectors which were both simpler and cheaper.

British railways in particular were loath to make use of patented devices from outside suppliers or equipment on which patent royalties had to be paid. They had a healthy scepticism regarding the savings claimed for these products. None the less, the LNER gave a full and fair trial to the ACFI type of feed-water heater. This was fitted to two ex-North Eastern Atlantics and then, from 1927, to 50 or so ex-Great Eastern 'B12' 4-6-0s which were transferred to the former Great North of Scotland section. Whether it was because these engines were far from their original home, or because their boiler-tops were burdened with the ACFI equipment, or because the fireman faced a longer than customary distance between tender and firedoor is uncertain but the relocated 'B12s' gained the nickname of 'Hikers'. Two Gresley Pacifics and the pioneer 'P2' 2-8-2 No 2001 *Cock o' the North* also underwent trials with the ACFI-pattern heaters, but by the late 1930s the LNER apparently concluded that they were not worth the trouble of extra maintenance.

By the 1930s, some significant changes were being made to some pre-Grouping designs, usually to increase either their usefulness or reduce their operating costs. As with the feed-water devices, the claimed savings achieved, for instance, by use of poppet valves operated by rotating cams were not always attained in everyday service. The same was true of the 'boosters' or auxiliary engines fitted to ex-GNR and NER Atlantics and 0-8-4Ts of ex-GCR design, as well as one of the pair of Gresley 'P1' 2-8-2s.

Locomotive boilers tended to have a life of 20-25 years, depending on the standard of maintenance and the relative hardness of water in the area in which the engines operated. It made sense for companies to try to reduce the number of types of boilers used. To that end, from the early 1930s four classes of ex-Great Eastern engines were chosen to receive boilers of

Gresley, round-top design. These were the 'J18' and 'J19' 0-6-0s, all rebuilt to Class 'J19/2' from 1934-39; the 'B12' 4-6-0s, 54 of which were altered to part 'B12/3' between 1932-44; and the 'D16', or 'Claud Hamilton' 4-4-0s. Of the last type, out of the class total of 121 all but 17 were redesignated 'D16/3' as a consequence of their rebuilding over the period 1933-49.

Another rebuilding project involved the 45 Raven 4-4-4Ts (LNER 'H1' class) of the North Eastern Railway. They had been built to work local passenger trains on Teesside and Tyneside but seemed not to be very popular. In the early 1930s, summertime traffic on the coastal lines serving Scarborough, Whitby and Middlesbrough was growing and the LNER required more motive power. It needed engines capable of keeping their feet on days when rails were coated by sea-spray or dew. To fit the bill, from 1931 all members of the class were rebuilt as 4-6-2Ts of Class 'A8'. The proportion of adhesive weight available had been increased with the addition of another driving axle.

The three-cylinder Raven mixed traffic 4-6-0s of LNER Class 'B16' were another choice for reboilering. Solid and reliable though these engines might be, it

Below:
A GER 'Claud Hamilton' 4-4-0 which has been rebuilt with Gresley round-topped boiler as Class 'D16/3'. No 8821 is working a down outer suburban train past Chadwell Heath on the Colchester main line in June 1933. No 8821 had been recently rebuilt and was to last until 1958 as BR No 62572. *E. R. Wethersett*

was thought that they might benefit from extensive re-engineering that involved fitting a new front end with the cylinders cast integral with the smokebox saddle and steam chests. Walschaerts' valve gear with Gresley-derived drive replaced the original Stephenson link motion on the seven engines which in altered form became Class 'B16/2'. Whether the cost of such alterations was worth while must be debatable. One questionable rebuilding of the time was of ex-NER 'D20' 4-4-0 No 2020 which was fitted with long-travel valves and otherwise altered to the directions of Edward Thompson. Significantly, no other 'D20s' followed suit.

Other ex-NER classes such as the Atlantics and the various classes of goods engines faithfully served the LNER, and then BR, with little more than minor changes to their anatomy although, like all steam locomotives, they had their fair share of boiler swaps or brand-new boilers and replacement cylinders fitted.

Some say that Gresley was open to new ideas, from whatever direction. That was true - but only up to a point. It had taken very great persuasion from his immediate colleagues to get him to fit long-travel valves to the 'A1' Pacifics after the trial runs on the LNER with GWR No 4079 *Pendennis Castle* in 1925 had highlighted the limitations of the 'A1s' in their original form. From 1927, these engines were fitted with boilers pressed to 220lb pressure and new

engines were so constructed from 1928 as Class 'A3'. Cylinder lubrication was improved for the engines with long-travel valves. Later boilers had banjo-shaped domes (indicating that a new type of steam collector was fitted) and new engines incorporating all the improvements were referred to as 'Super Pacifics'. The last to be built were Nos 2500-2508 outshopped in 1934/5. With trains of up to 600 tons gross, the 'A3s' frequently demonstrated their ability to run at over 60mph on level or slightly rising gradients.

Those East Coast and West Riding expresses of 65 years ago were almost entirely made up of Gresley's teak-panelled corridor coaches of both pre-Grouping and more recent construction. They looked remarkably similar, as the basic shape and appearance that dated back to 1905 was maintained until the early years of World War 2.

Some of the coaches in the expresses were articulated, mainly the restaurant car triplets used on Anglo-Scottish, Newcastle and Leeds trains. Articulation, said Gresley, saved on first cost, and reduced the length and weight of trains. Less exalted

Below:
Ex-North British Railway 'Intermediate' class (LNER 'D32/2') 4-4-0 No 9885 passes Haymarket locomotive depot with an Edinburgh Waverley-Arbroath train, c1932. The first half of the train is made up of standard LNER non-corridor stock. No 9885 was withdrawn in 1948. *G. R. Grigs/Photomatic/Rail Archive Stephenson*

Above:
Just two days to go before British railways pass into Government control with the approach of war. The date is 28 August 1939 and 'V2' 2-6-2 No 4825 lays a smokescreen over Sandy as it heads towards King's Cross with an up express.
E. R. Wethersett

outer suburban and suburban stock was also articulated, sometimes in four-car units (five-car on the GER suburban lines). By the mid-1930s, another species was appearing, with the articulated twins of steel-panelled corridor coaches that made up trains for services to Peterborough and Grimsby.

The triplet restaurant cars worked in the 'Flying Scotsman' and the 'Afternoon Scotsman' departures from King's Cross and Edinburgh. From 1928, the 'Flying Scotsman' had run nonstop between the capitals in summer only, and the triplet restaurant cars had included a first-class car with Louis XVI styled interior. On the East Coast route, nonstop running was possible - a publicity gimmick, and a powerful one - because of a well-spaced series of water-troughs and the use of corridor tenders with the Gresley Pacifics that enabled engine crews to change over without the train having to come to a stand. They walked from the train through the tender and on to the footplate. The two daytime East Coast expresses in each direction were accorded priority by the LNER. Not far behind were other trains such as the 'Breakfast Flyer' from Leeds to King's Cross which had been the first train to be put on a fast timing, the 5.30pm King's Cross-Newcastle, and the Pullman car expresses, such as the 'Queen of Scots', 'West Riding' and Tees-Tyne Pullman'.

By the end of the 1930s, the Gresley Pacifics had been joined by the capable 'V2' 2-6-2s, the class-leader of which was named after the Green Arrow service to freight traders mentioned earlier in this section. The 'V2s' were at home on all sorts of working, and particularly the 'flying' fitted freights.

The quickening pace for GNR main line expresses dated from 1932 then, two years later, came *Flying Scotsman's* famed run to Leeds and back, giving a clear sign of the sort of speeds - 100mph - a Pacific was capable of at the head of a light train. Just a year later, 'A3' *Papyrus* ran to Newcastle and back, reaching 108mph on the return down Stoke Bank. That second run confirmed the LNER's top brass in the wisdom of introducing a high-speed steam-hauled express between King's Cross and Newcastle.

In no more than six months Gresley and Doncaster Works produced a complete train of articulated coaches with a striking finish of silver-grey fabric covering steel panels and trimmed in stainless steel; with the streamlined 'A4' Pacific No 2509 *Silver Link* at its head. The introductory run of the 'Silver Jubilee', as the new train was called in honour of King George V's silver jubilee year, took place on 27 September

1935. This was 110 years after the opening of the Stockton & Darlington Railway. No matter that the maximum speed was 112mph on this special run, more impressive was the fact that no less than 43 miles on the outward journey to Grantham had been covered at 100mph. The LNER reckoned that the introduction of the 'Silver Jubilee' acted as a stimulus to traffic which increased by 12% between Newcastle and King's Cross.

The 'Coronation' of 1937 was the best paying of the LNER's three high-speed expresses, and the train sets were carefully designed with the aim of permitting at-seat service of drinks and meals to *all* passengers. The articulated twin-sets making up the main eight coaches had a centre gangway layout and were painted in a two-tone blue paintwork that was kept as well-polished as any limousine. In summer, an observation car was at the rear of the train. In its first year of service from 1937, the 'Coronation' earned a profit of 13s 8d per loaded train mile. The other LNER high-speed trains were the 'West Riding Limited' - King's Cross-Leeds/Bradford - with almost identical stock as the 'Coronation', and the 'East Anglian', from Norwich to Liverpool Street and back. The 'East Anglian' was not high-speed when it came to timings but the principle of on-board service was the same as the 'Coronation' although it proved to be half as profitable as the 'Coronation'. With their teak-panelled exteriors, the 'East Anglian' coaches were no different in appearance to standard Gresley stock.

These high-speed trains were operated very professionally but one suspects that Gresley wanted to see for himself what an 'A4' could do when there were no fare-paying passengers aboard. The chance was taken in July 1938, at the conclusion of a series of weekly braking trials. With 'A4' No 4468 *Mallard* at its head, a set of streamlined coaches and the LNER dynamometer car briefly reached 126mph down Stoke bank. The high-speed era on the LNER had reached its pinnacle.

No more than a year's operation remained for the 'Silver Jubilee', 'Coronation' and 'West Riding Limited'. Their final day was 31 August 1939. The next day, the Big Four passed to overall Government control again and emergency services had priority. On that fateful Thursday, the very last southbound 'Coronation' left Edinburgh Waverley at 4.30pm behind 'A4' No 4488 *Union of South Africa*. No more than 74 passengers were aboard, 20 of whom got off at Newcastle where 58 boarded for King's Cross which was reached one minute late. This was the sad but proud finale - not only of the LNER's streamliners but also the spirit of our interwar railways.

Below:
The up 'Silver Jubilee' has left Newcastle Central at 10am and is seen near Durham, c1938/9. The locomotive is blue-liveried 'A4' Pacific No 2509 *Silver Link* and the eight-coach 'Silver Jubilee' set is spray-painted silver-grey. *Colling Turner/Photomatic/ Rail Archive Stephenson*

4. **The 1940s:**

*Wartime Control to Nationalised Unity - Evacuation
Trains to Faceless Bureaucrats*

The Big Four companies had been planning for their role in war for at least two years before hostilities commenced. Committees sat to consider the conversion of passenger stock to form ambulance trains, what might be required in the way of diversionary routes, and examined all aspects of air raid precautions.

The railways passed to Government control on 1 September 1939. That was the day when a mass three-day evacuation began of women and children, very largely travelling by train, from London and other major cities. From southeast London alone, 225 special trains conveyed over 100,000 people to country stations; not surprisingly, at times the normal train service was curtailed. The Government's belief had been that with the onset of war extensive air-raids on cities would follow speedily and that, to avoid chaos and panic, the most vulnerable must be sent away in advance.

Soon after the declaration of war on 3 September, it became clear that the much feared destruction of the cities would not take place, at least for the time being. Britain moved instead into the so-called 'Phoney War' and train services settled down.

From 25 September 1939, the GWR had brought in its first wartime timetable, with widespread curtailments. The remaining trains were decelerated - all expresses were restricted to a start-to-stop average speed of 45mph. Until 11 September, the SR had continued to work normal summer services (save for the evacuation trains), but thereafter services were curtailed and decelerated, only to be partially relaxed the following week and cut back again soon afterwards. From late 1939, there were no midday services on the Dover, West of England and Bournemouth lines so as to allow troop and freight

Above:
At first sight, this is a normal station scene but it is wartime and the newspaper billboards read 'Latest War News' and the various road vehicles have white-painted details to reduce after-dark hazards. The location is Watford Junction station forecourt.
Science & Society Picture Library LMS 9299

services to run. Loads on the Bournemouth line, for instance, increased to 16/17 coaches.

On the LMS, normal summer services continued to the second week of September, apart from the 'Coronation Scot' which had been withdrawn at the outbreak of war. Like the GWR and LNER, the war emergency timetables applying from late September featured services that were slower and less frequent, particularly on the Midland Division where the average booked time from St Pancras-Manchester was as much as 6hr.

The first wartime timetables on the LNER came into force in October 1939 when the East Coast Scottish day trains were reduced to just one each way, and the GCR main line services were drastically cut back. Before long, East Coast services were supplemented by virtually regular relief trains and principal services were loading to 20 coaches. Yet, in May 1940, the scheduled main line service out of King's Cross was half that of prewar days.

Once the likelihood of invasion had receded, some main line services were reinstated during the summer of 1941, particularly to/from the West Country, and speed limits were relaxed. Wartime passenger train services were always open to change at short notice. When it was thought important to free the South Wales main lines for additional coal trains the passenger service from Paddington and Bristol to South Wales was temporarily suspended. During 1941, to provide more paths for military specials, some of the Southern Railway's West of England services were

combined, and then loaded to as many as 20 coaches, on which Bulleid's new 'Merchant Navies' showed their mettle. Despite attempts to curb recreational travel, people were not dissuaded, and in 1942 summer long-distance traffic on the LMS was some 74% above that for 1938.

The advertised wartime service was no clue as to the number of trains actually run. One Friday night in September 1942, of seven expresses arriving at King's Cross within half an hour, three were the main trains, the others their duplicates; in total they comprised 132 coaches. Not all the principal trains had been withdrawn on the outbreak of war and the 'Aberdonian' and 'Night Scotsman', for example, turned out to be two of only four trains that retained their titles throughout hostilities.

Overall, working conditions for railway staff were very difficult while the travelling public endured some horrendous journeys, with people packed into trains and numerous delays and alterations. Once the Luftwaffe began its air raids, train journeys were seriously affected by alerts. At night, blackout regulations made railway stations and on-train conditions irksome and eerie. That was nothing compared to the difficulties faced by enginemen on locomotive footplates with the anti-glare curtains making their work both arduous and oppressively hot.

After the war, the four railway companies published accounts of the wartime experience. They paid suitable tribute to the heroic deeds, sheer hard work and stoicism of their staff, 385 of whom lost their lives as a result of enemy attacks on the railways. Enginemen on exposed footplates frequently faced chases from low-flying aircraft and cannon fire.

The work involved in reinstating sections of line following severe air attack was sometimes beyond belief. On the Southern Railway alone, in nearly five years of aerial attack 14 bridges and viaduct arches were demolished by bombing, and over 180 suffered varying degrees of damage. Air raids sometimes took out whole sections of railways, and more than one station at a time. On the morning of 17 April 1941, five of the Southern Railway's London termini were closed to traffic: Waterloo, Victoria, Charing Cross, London Bridge and Holborn Viaduct. Burning buildings almost surrounded Waterloo station where

Above:
Initial attempts to restrict train interior lighting were disastrous for passengers and visibility was greatly reduced. This is the 'improved lighting' introduced by the Southern Railway later in 1939. *Ian Allan Library*

Below:
A huge bomb crater, torn-up track and damaged buildings beside the North London line near Highbury. An LMS Class '5' stands clear of a train being loaded with debris after an air raid. *Ian Allan Library*

all utilities supplying the station had been put out of action by bombing.

Huge reliance was placed on the railways for the movement of military hardware, stores and troops. Military freight trains had preference over all other freight services, except for fully fitted freights. The programme of airfield construction, particularly in the Eastern Counties, called for the transport of steel, cement and other building materials, and once the airfields had been completed the railways had to supply them with fuel.

Collectively, the railways were presented with huge challenges. Just before Dunkirk in 1940, the Big Four general managers were called to a meeting with Winston Churchill at the Charing Cross Hotel to plan the evacuation of British troops by train from the Channel Ports. Having heard from Churchill what was involved, someone who attended the meeting reported that one of the GMs made the immortal comment - 'It's just another FA cup final!'

Some 335,000 allied troops were rescued from Dunkirk in May 1940 (Southern Railway shipping in particular played a noble and vital role in the operation) and, over a period of 16 days from 27 May, they were dispersed from the South Coast of England using no less than 186 sets of coaches which worked a total of 620 specials. Such organisation was conducted through the train and traffic control offices, most of which were relocated into what was hoped were 'safe' areas. In heavily reinforced offices staff toiled endlessly, many working 20 out of 21 days before taking a brief respite, all the time being kept aware of the position regarding air raid alerts and incidents.

By late 1942, the railways had become very congested with traffic movements, particularly as some heavily used but ill-equipped routes lacked the required capacity. To improve matters, new junctions between routes, additional refuge loops, and miles of sidings were installed.

The lead-up to the Allied forced landings in Europe on 6 June 1944 (D-Day) was preceded by extensive train movements. In the two previous months as many as 22,000 special trains conveyed troops across the country. The movement of military stores and equipment for D-Day had taken place in advance and entailed the operation of 800 special trains; in addition, thousands of individual wagons were moved by scheduled freight services.

Below:
Troops evacuated from Dunkirk are served refreshments at Kensington Addison Road station. *Imperial War Museum*

Throughout the war, the railway workshops were engaged in military production, quite apart from their usual duties. They turned out a vast array of munitions and equipment for all three armed services. All the time, the tally of destroyed vehicles was lengthening, not to mention arrears of maintenance that would have to be recouped after the war ended. Of the combined Big Four fleets, although just eight locomotives were written off, 637 coaches and 2,685 wagons (including private owner wagons) were destroyed.

By 1944 at least, the managements of the Big Four had begun to look at what might be done with their part of the railway system after the war. In August 1944, the Southern Railway set up a committee to study and report on the completion of electrification of the Eastern and Central Sections and the report was ready in February 1946. It recommended the completion of electrification at a cost of £11.9 million and that the scheme should be effected over a period of not less than ten years. The GWR contemplated a wide range of improvements including the

Above:
In with the new! The British Railways number is applied to the cabside of an LMS '5' 4-6-0 at Crewe Works. *Ian Allan Library*

Right:
Wartime grime did not suit the streamlined LMS Pacifics. No 6240 *City of Coventry* is cleaned at a late stage in the war. *Ian Allan Library*

introduction of gas turbine locomotives, fluorescently-lit coaches and an automated buffet car, and published a book outlining its ideas.

The LNER set up a Post War Development Committee as early as 1942 and this reported to the Board early in 1944. The Committee's report envisaged the electrification of main lines from London, on the grounds that steam would probably be prohibited altogether within a radius of 25 miles of the capital. New London termini were contemplated, as was the electrification of suburban lines on Tyneside and Glasgow, and completion of the Manchester-Sheffield scheme.

Such plans were forestalled by the election in July 1945 of a Labour Government whose election programme had included a demand for public ownership of the railways, although during wartime even Government ministers had advocated that the industry should remain under central control for the foreseeable future. Before the end of 1945 the new Government announced its intention to introduce measures to bring transport services under public ownership and a Transport Bill outlining the proposals was published in the autumn of 1946. The Bill gained Royal Assent in the August of the following year and the British Transport Commission began its operations on New Year's Day, 1948. The Big Four companies had lobbied hard but unsuccessfully against nationalisation.

The British Transport Commission was just what the name implied - an overall transport authority which would own and control public inland transport although only part of this ambition was realised. The largest of its functions was the Railway Executive which managed the railways. After some other proposals were considered, the railway system was divided into six Regions - Eastern, London Midland, North Eastern, Scottish, Southern and Western.

It took some time for the new organisation to establish itself, and in the meantime the public continued to endure some of the ills of the railway system that had been all too common during wartime. Although the Big Four had genuinely tried to return quickly to some semblance of prewar services, by the summer of 1947 the improvements achieved were reversed and the standard and levels of service slipped downhill, not least because of a shortage of coal. There was also a backlog of locomotive and rolling stock repairs and, all the while, the condition of the way and works and public facilities continued to deteriorate. The Southern Railway admitted that 890 of its steam locomotives were 40 or more years old. The LNER estimated that because the war had caused their withdrawal to be deferred, 2,500 over-age coaches had been kept in service. There was a shortage of railway workers and their rates of pay fell steadily behind other industries.

At first nationalisation seemed to involve the creation of centralised committees, deferment of existing modernisation schemes, and an emphasis on publicising the name of the new undertaking. There were allegations that faceless bureaucrats were directing railway policy. In fact, productivity improved steadily from 1948, and the 285 million tons of freight carried in 1951 constituted a postwar record, moved by using 1,000 fewer engines than in 1948. Wagon productivity improved and, although the run-of-the-mill freight train continued to be devoid of through braking, the number of scheduled vacuum-braked express freight trains was increased.

BR claimed that it ran the busiest railways in the world and that its trains covered a greater mileage than any other European country. For 1948 and 1949 there was a surplus of receipts over expenditure as indeed there was into the early 1950s. Under BR the existing locomotive and rolling stock fleets were rationalised. Apart from the reduction in the steam locomotive fleet, the numbers of wagons in service also diminished. In 1948 BR operated no less than 1.179 million wagons, of which 489,000 had been private-owner wagons (ie, not owned by the railway companies) but by 1950 the total was 1.104 million. Over 100,000 ex-private wagons had been taken out of traffic.

Soon after nationalisation, the Railway Executive set up committees to produce standard ranges of steam locomotives, coaches and wagons. They worked remarkably quickly although the shortage of steel meant that the intended production rates were not

Left:
One of the 2-8-0s built for the War Department and loaned to the British railways passes West Brompton with a southbound goods on 15 April 1950. This engine, WD No 77030, was later taken into BR stock as No 90127. *E. R. Wethersett*

Below:
Platforms 14 and 15 at Euston station, c1946/7. At the time the conditions would not have been thought exceptional but dingy and smoke-filled station interiors added to public perception that railways were a less than desirable way to travel. *Ian Allan Library*

achieved. Much was made of the fact that BR was operating over 400 different types of steam locomotives and that eventually new production would be covered by 12 or so Standard designs. These would be able to operate almost throughout the system, it was claimed, and would be both cheaper to maintain and also have certain interchangeable fittings and components. The first Standard engines appeared in 1951. In the interim - and indeed for a year or two longer - the production of some existing Big Four designs continued. The first Standard coaches also appeared in 1951.

Despite some undoubted developments the public perception of railways was poor, no doubt influenced by a popular press which was generally hostile to nationalised industries. In railway terms, the 1940s would be recalled as a decade of austerity, of overcrowded, often grubby and worn-out stations and rolling stock and of cancelled trains.

This set of images was fired into the public consciousness more than any new style that BR might try to establish. The new liveries adopted for locomotives and rolling stock were confirmed in 1949 following trials of various strange and sometimes unsuitable combinations of colours and lining-out. The standard liveries saw the most prestigious express passenger locomotives painted a darkish blue, lined-out in black and white; all other express passenger locomotives had dark green paintwork, lined out in black and orange. This attractive scheme soon replaced the blue livery. Other passenger and mixed traffic engines were painted black, lined-out in red, cream and grey, and freight and shunting engines were plain black. Express passenger coaches were painted crimson lake with cream panels, lined-out in gold and black. All other coaches were plain crimson lake (this shade was also referred to as carmine) and electric trains were painted unlined green. Unfitted wagons were painted light grey, most fitted wagons were bauxite. A new numbering system allotted ranges of numbers for each of the Big Four companies, and for diesel and electric locomotives.

Each of the BR Regions adopted a colour to identify its station nameboards, direction signs, the cap badges of staff and for its timetables and much printed material. The Eastern Region used dark blue; the London Midland, maroon; the

North Eastern, tangerine; the Scottish, light blue; the Southern, green; and the Western Region, chocolate. All railway ships had black hulls with a white superstructure, and buff funnels with a black band at the top.

From GWR to WR in the 1940s

Superficially, the GWR had changed little in the last 10-15 years, certainly on the rural routes that had escaped bomb damage and alterations to their way and works to accommodate wartime traffic flows. The single-track Didcot, Newbury and Southampton line, for instance, was partially doubled and crossing loops were installed or lengthened on its southern portion to make it easier to handle freight trains travelling to and from Southampton Docks.

By contrast, a 1940s' journey from Moreton-in-Marsh to Banbury via Kingham would have revealed that the main intermediate stations and wayside halts had

Right:
Pair horse van with high dicky-seat at King's Cross Goods depot. After the war, the Eastern Region moved quickly to mechanise cartage from its London stations.
Ian Allan Library

hardly changed from prewar days. The main development dated from 1929 when the branch from Moreton-in-Marsh to Shipston on Stour had lost its passenger service, one of the earlier such withdrawals. Some of the services along parts of the Oxford-Worcester line were worked by GWR diesel railcars working from Leamington Spa to Evesham, Oxford to Kingham and Chipping Norton. The Cheltenham Spa-Kingham-Chipping Norton-Banbury line had been traversed by the Newcastle-Swansea through express which had been one of the first workings to be covered regularly by 'Manors' when these had been new. The service did not return to the cross-country line after World War 2.

Banbury station had opened in 1850 and was notable for its overall roof. It had survived more or less unchanged except for additional offices provided in 1910 when the Bicester route had come into use. By the early postwar years the condition of Banbury station's overall roof was such that drivers were warned not to let their engines blow-off underneath the frail wood and glass structure. The overall roof was removed in 1952 and three years later work started on Banbury's new station which was completed in 1959.

Banbury was a major centre on the railway network during World War 2, principally on account of heavy traffic passing over the Great Central main line and then via the Culworth Junction-Banbury link. The lever frame at Banbury South signalbox was extended to cope with additional train movements, the nearby marshalling yard enlarged, the goods shed renewed, and facilities at the loco shed augmented and improved.

Mention was made earlier of the GWR railcars, of which 38 had entered traffic between 1934 and 1942. These were the predecessors of the many diesel-mechanical railcars that were put into service by BR under the 1950s' Modernisation Plan. Two of the GWR railcars were loaned to the LNER during the war to run trials in the Newcastle area, and in 1946 the LNER worked out an unrealised scheme to introduce diesel-mechanical railcars in the Tyne and Wear areas, and working also from Leeds and Hull.

The first of a new series of GWR locomotive designs appeared during World War 2, after F. W. Hawksworth had taken over from Charles Collett as Chief Mechanical Engineer. Numbered upwards from 6959, the 'Modified Halls' began to be turned out by Swindon Works during 1944. They marked the first

Below:
A Swansea-Paddington express, east of Twyford, behind No 5099 *Compton Castle*, the second-built of the postwar 'Castles' which were fitted with increased superheating and mechanical lubricators. No 5099's tender is lettered British Railways. *R. F. Dearden/Ian Allan Library*

changes from design features that had been incorporated in Churchward standard classes, notably by the move to a plate-frame leading bogie, changes to other mechanical parts to ease manufacture, and provision of a larger superheater. A year or so later came an entirely new design of two-cylinder 4-6-0, the 'County' class. These introduced a new boiler type with a working pressure of 280lb whose manufacture owed something to the design of the boiler used for the Stanier '8F' 2-8-0s built at Swindon during the war. Other innovations were the flat-sided tenders used for the 'Counties' and the 6ft 3in driving wheels, a new size for the GWR. The class leader, No 1000, uniquely carried a double chimney. Both the 'Modified Halls' and the 'Counties' had increased superheating and the boiler type used for the 'Castles' was similarly altered when new construction of this class resumed in 1946. The later postwar 'Castles' also had mechanical lubricators in place of the sight-feed type. There was much debate about the possible emergence of a Hawksworth Pacific. Evidence suggests that it amounted to no more than a drawing office sketch.

The other new GWR locomotive designs were smaller engines. In the year before Nationalisation a new class of pannier tank was introduced - the '94xx' class - with taper boiler and wider cab than the '57xx' class. Less conventional was the '15xx' class, 10 of which appeared after nationalisation. These shunting engines were unsuperheated, had outside cylinders and Walschaerts valve gear and a short wheelbase for working in goods yards; however, at least four of the class spent most of their lives hauling empty coaching stock to and from Paddington. The last new GWR-pattern design to emerge from Swindon was the '16xx' class 0-6-0PTs which were effectively an updated version of the venerable '2021' class that dated from Victorian days. The '16xx' class proved useful for mixed traffic duties on lightly-laid branch lines, and for shunting.

For its passenger rolling stock, the GWR moved to an altogether new design with slab-sided corridor coaches with sloping ends to their roofs. Although these vehicles were of relatively conventional construction, the railway experimented with aluminium alloy construction, with the use of laminated plastics for interior panelling, and with

Below:
One of the official photographs of newly completed '15xx' 0-6-0PT No 1500. *Ian Allan Library*

Left:
In addition to new designs, construction also continued of earlier GWR standard classes, such as this '57xx' 0-6-0PT, No 9643, which was photographed when under construction at Swindon Works in April 1946. *H. C. Casserley*

Right:
On its way out - 'Bulldog' 4-4-0 No 3417 (*Lord Mildmay of Flete* and formerly *Francis Mildmay*) shunts in the goods yard at Princes Risborough in April 1948, the month in which the veteran 4-4-0 was withdrawn from service. *H. K. Harman/ Rail Archive Stephenson*

Left:
On 3 August 1947, oil-burning No 3952 *Norcliffe Hall* heads the 11.15am Sundays Paddington-Bristol-Taunton express near Durston, on the approaches to Taunton. On conversion back to burning coal, the engine reverted to its former number of 6957. *Pursey C. Short*

fluorescent lighting. Construction of this so-called Hawksworth stock took place from 1947-51.

With the arrival of new classes of GWR locomotives, combined with the continuing construction of well-established types such as the 'Manor' 4-6-0s, '2251' 0-6-0s, '5101' 2-6-2Ts, and '57xx' and '74xx' 0-6-0PTs, many older engines were withdrawn from service. Classes eliminated or suffering heavy withdrawals included various classes of tank engine from the South Wales railways, 19th century 0-6-0PTs, 'Aberdare' 2-6-0s, 'Saint' 4-6-0s, 'Bulldog' 4-4-0s, 'Metro' 2-4-0Ts and '517' 0-4-2Ts.

In the autumn of 1945 the GWR embarked on a scheme to convert a number of its locomotives in South Wales as oil-burners, a response to controls imposed by Government on the allocation of coal supplies. The project got under way in October 1945 when a '28xx' 2-8-0 entered service as an oil-burner, the first of 10 of this class considered for conversion. All were based either at Llanelli or Severn Tunnel Junction sheds where oil-fuelling facilities were installed. Early in 1946, it was decided to convert one each of 'Castle' and 'Hall' class 4-6-0s. That May the company decided to equip 25 'Castles' in all, and to build oil storage

depots at seven sheds around the system. This project was financed by the Company and carried out to its specifications.

Before long, the GWR's project became embroiled in a much greater scheme. Early in June 1946, the Ministry of Fuel and Power approached the wartime and early postwar Railway Executive Committee to enlist its help by converting steam locomotives to burn oil. The GWR's earliest oil-burner conversions used Swindon-designed equipment and a modified installation was accordingly used as the basis for the oil-burning conversion of all engines in the Government scheme.

The national scheme to convert steam locomotives as oil-burners got out of control, once Whitehall decided to force the railways' hands. Experience with oil-fired engines was that the maintenance of boilers and fireboxes was increased, and overall fuel efficiency when burning oil was demonstrably less than with coal.

Below:
The country end of Southampton Central station with the down 'Bournemouth Belle' Pullman car express. 'N15' 4-6-0 No 755 *The Red Knight* is alongside the platform end and nearer the cameraman is ex-LSWR '0395' class 0-6-0 No 3101.
Frank F. Moss

The SR in the 1940s

Some people, particularly outside southern England, thought of the Southern Railway as largely electrified by the late 1940s. This was far from the truth and the Eastern Section's main lines beyond London's suburban area, and all but the Portsmouth line and suburban routes of the Western Section were steam-worked. Even the Central Section had the steam-worked Oxted lines and one or two other routes. Then there was the former South Eastern Railway's line from Reading via Wokingham, Guildford, Redhill and Tonbridge which for part of the way ran past the picturesque North Downs. Its services were steam-worked although they traversed short sections of electrified line, such as Reading to Wokingham, Ash to Guildford, and Reigate to Redhill. The railway geography in the Farnborough/Aldershot area was - and remains - complicated, with the Reading-Tonbridge route threading through the former London & South Western network of lines to and from Waterloo and Alton and Guildford. Between Ash and Guildford, and out of Guildford to Shalford Junction, the Reading-Tonbridge trains ran over LSWR metals. In the immediate Aldershot area there were several

stations, including North Camp which the timetable helpfully explained in its footnote F was 'Station for Ash Vale and South Farnborough'.

In the last summer of the Southern Railway, trains on the Reading-Tonbridge axis - part of which was the former South Eastern Railway's London-Redhill-Dover main line that ran as straight as a die through East Surrey and Kent - functioned as two services, between Reading and Redhill, and Redhill and Tonbridge. The exceptions were early morning through trains from Reading to Margate, and from Reading to London Bridge via Redhill, the latter being for season ticket-holders from North Downs stations. Still missing from the August 1947 timetable were the long-established cross-country trains in each direction between Birkenhead Woodside and Dover/Ramsgate.

If we return to North Camp station, it was served by all the Reading-Guildford/Redhill stopping trains. Despite the fact that for most of the Southern Railway period the Reading-Guildford-Redhill service came under the Central Section, it remained a refuge for

Below:
A Reading-Redhill line train - ex-SE&CR 'D' 4-4-0 No 31574 - at Reigate with an electric train recently arrived from Redhill alongside. *Ian Allan Library*

South Eastern engines such as the 'F1' 4-4-0s. For many years the various classes of Maunsell 2-6-0s were used for the line's passenger trains although ex-SE&CR 'H' 0-4-4Ts, ex-LSWR 'M7' 0-4-4Ts and ex-LBSCR 0-6-2Ts were used on some workings west of Redhill. Ex-LSWR 'K10' 4-4-0s were also employed on some Reading branch trains, as were ex-SE&CR 'D' class 4-4-0s.

The Birkenhead-Margate through train, with through coaches for Hastings via Eastbourne and Brighton, reappeared in the timetables of the late 1940s. By the winter of 1949 it boasted a through refreshment car in the Margate section and, leaving Birkenhead at 7.40am, pulled into Margate at 4.53pm. That was not all: having detached the Hastings portion at Redhill, the train divided again at Ashford, the main section with refreshment car running via Canterbury West, and the other portion bound for Ramsgate via Dover and Deal. This procedure was followed in the reverse direction.

Enthusiasts were always interested in cross-country expresses such as these. Another well-known service was that from York to Bournemouth via Sheffield, the ex-Great Central main line and Oxford, and the Birkenhead-Bournemouth train, routed via Shrewsbury and Birmingham Snow Hill. These services had restaurant cars throughout their journeys. Other cross-country trains travelling over Southern Railway routes for parts of their journeys were those between Reading and Portsmouth, and between Cardiff, Bristol, Salisbury and Portsmouth/Brighton. The Reading-Portsmouth workings were generally handled by ex-GWR motive power. Then there was one cross-country express that ran almost entirely over SR track - the Brighton/Portsmouth-Salisbury-Exeter-Plymouth train.

One vantage point for some of these through trains was Basingstoke where the Great Western line from Reading joined the South Western main line. Through Basingstoke there were four tracks until, just under three miles west of Basingstoke, the South Western's West of England and Bournemouth main lines diverged at Worting Junction, just west of which at Battledown a flyover took the up Bournemouth line over the West of England lines in order to eliminate conflicting movements. Basingstoke station was well laid out with two island platforms (and adjoining separate Great Western platforms). From 1901-17 and

Below:
Cross-country express - the Birkenhead-Margate train passes Ash Junction on 7 September 1949 behind 'U' 2-6-0 No 31798 which is at the head of ex-GWR coaching stock. *E. C. Griffith*

1924-36 Basingstoke was the junction for the Basingstoke & Alton Light Railway. The gap in dates is because the line was closed during World War 1 and only reluctantly reopened by the Southern Railway. Complete closure came in 1936.

All sorts of locomotives came sweeping through Basingstoke on the main lines. By the late 1940s these included Bulleid Pacifics of the 'Merchant Navy', 'West Country' and 'Battle of Britain' classes. Despite the numbers of the new air-smoothed Pacifics, the older Maunsell 4-6-0 express locomotives of the 'Lord Nelson' and 'King Arthur' classes were well in evidence, backed up by ex-LSWR mixed traffic 4-6-0s of the 'H15' and 'S15' classes, and one or two less numerous types, including the last survivors of the Drummond 'T14' or 'Paddlebox' 4-6-0s which just lasted into the 1950s.

On summer weekends nearly all the classes just mentioned would be at work on holiday trains bound for Bournemouth, Weymouth and Swanage, Ilfracombe, Padstow and Bude, Sidmouth and Exmouth. That was just the service to and from Waterloo. There were also the regular cross-country expresses, supplemented by additional holiday peak trains between the North,

Above:
Basingstoke - on 25 April 1950, a Waterloo-Salisbury slow train departs behind 'King Arthur' 4-6-0 No 30772 *Sir Percivale. E. C. Griffith*

Midlands and Portsmouth and Bournemouth. Bournemouth was of course also served by through expresses routed over the Somerset & Dorset line.

One feature of weekdays were the numerous boat trains between Waterloo and Southampton Docks, freight trains including banana specials between Southampton Docks and Nine Elms, and others bound for Feltham marshalling yard. These were largely in the hands of the various classes of ex-LSWR and SR mixed traffic 4-6-0s, and at summer weekends these engines were redeployed either on expresses or on semi-fast trains, in turn releasing 'King Arthurs' for holiday trains. Some 4-4-0s remained at work and one speciality of the immediate postwar years was the use of the Drummond 'D15' 4-4-0s on the through Saturday holiday trains between Waterloo and Lymington Pier.

The advent of the Bulleid Pacifics had relegated the Maunsell 4-6-0s to less important trains, at least during the week. The style of the Southern's Western Section

in these years on either side of nationalisation was set by the combination of the Bulleid Pacifics and the designer's handsome corridor stock on principal workings between Waterloo and Bournemouth/Weymouth, and Waterloo and the West of England. The restaurant cars included the Bulleid tavern cars with imitation brickwork, timbered beams and inn-signs painted on their outsides. The tavern cars were largely used on the West of England expresses, and conventional dining cars on the Bournemouth line trains. The real celebrities of the Western Section were the all-Pullman car 'Bournemouth Belle' and 'Devon Belle' expresses, both headed normally by 'Merchant Navy' Pacifics.

The Bournemouth train dated from 1931 and had been suspended during the war. It was reinstated on 7 October 1946, running weekdays and Sundays, Waterloo departure 12.30pm, Bournemouth West dep 7.15pm (altered to a 4.35pm departure the following June). The 'Belle' was given fast schedules, and for the 108 miles was allowed no more than 125min to Bournemouth Central, and the even 2hr coming up.

On its first run the 'Bournemouth Belle' was hauled both ways by 'Merchant Navy' Pacific No 21C18 *British India Line*. From its reintroduction, the train initially loaded to ten Pullmans, reduced to eight cars midweek. Within a year it was loading to 12 cars. The locomotive carried a rectangular headboard with the three-line inscription 'The Bournemouth Belle'.

The new Pullman train serving Ilfracombe and Plymouth was named the 'Devon Belle' and made its first public run on Friday, 20 June 1947 when in each direction the train was worked between Waterloo and the stop at Wilton South (where engines were changed) by 'Merchant Navy' No 21C15 *Rotterdam Lloyd*.

Running on Fridays to Mondays inclusive in its first season, the 'Belle' left Waterloo at noon. Its first stop at Wilton South was unadvertised and the first public call was at Sidmouth Junction. The train divided at Exeter Central, the first portion departing for stations to Plymouth Friary, the second departing for Ilfracombe. The two sets of coaches needed to work the 'Devon Belle' were formed of refurbished Pullman cars, each with an observation car at the rear of the Ilfracombe portion. The two observation cars were converted at Pullman's Preston Park Works from existing vehicles. In addition to their large bodyside windows, their outer ends were slightly raked with glazing extending from waist line to cant-rail and the cars' exterior paintwork was attractively styled. The cars each included a bar, pantry and lavatory and the saloon seated 27 in single- and double-tub seats.

Initially, 'Devon Belle' proved very popular and at times loaded to 14 cars, with 10 of these going through to Ilfracombe but after 1950 decline set in, the portion formerly running to Plymouth now travelling no further west than Exeter Central.

The third and fourth Pullman car expresses reintroduced by the Southern Railway after the war were the 'Golden Arrow' and 'Thanet Belle'. The 'Golden Arrow' was a long-established prestige boat train and began running again on 15 April 1946; in later years the outward run was from Victoria to Folkestone Harbour, returning however from Dover Marine. New features included a public address system fitted on board, a newly converted Pullman bar car, and at the front of the train would be a Bulleid 'Merchant Navy'; by the late 1940s a Light Pacific was more usual. A Pullman express to the Thanet coastal resorts dated back to 1921 but the 'Thanet Belle' of postwar years appeared later than

Opposite top:
The interior of one of Bulleid's tavern cars, showing the bar area. They were later converted to more conventional restaurant/buffet cars. *Ian Allan Library*

Opposite:
The down 'Devon Belle' makes an inspiring sight as it passes Hook Heath on 5 July 1947 behind 'Merchant Navy' Pacific No 21C18 *British India Line*. *Ian Allan Library*

Right:
One of the more unusual through trains on the Southern was the Yeovil-Dartford milk train, the empties of which are seen crossing the River Medway on their way to Tonbridge. There the train will reverse, to continue to Redhill (reverse again) and will reverse for a third time at Woking before heading west. The engine is ex-SE&CR rebuilt Stirling 'B1' 4-4-0 No 1448 which just survived into BR stock. *Norman Harvey*

Right:
The 'T9s' were still hard at work into the early BR years. This is No 302 at Mount Pleasant, Southampton, in April 1946 working a Portsmouth-Salisbury train, largely formed of GWR stock. No 302 lasted until 1952.
Frank F. Moss

Below:
The 'Leader' - No 36001 seen on a test train returning to Brighton.
Ian Allan Library

the other SR Pullman trains, as from May 1948, running as a daily train in summer only between Victoria and Ramsgate. Although it was hauled by Bulleid Light Pacifics, other engines such as 'King Arthurs' also appeared.

O. V. S. Bulleid, the Chief Mechanical Engineer of the Southern Railway, may have been responsible for comparatively few steam locomotive classes, but to his credit was the distinctive design of loco-hauled coaches and all-steel suburban electric multiple-units. Apart from the three Pacific classes there was also the 'Q1' 0-6-0 whose austere and almost unbelievable outline delivered such a shock to wartime Britain. Originally numbered C1-C40, the 'Q1s' were competent on all sorts of workings and possessed a surprising turn of speed. But they were conventional compared

with Bulleid's 'Leader' class two-bogie design. The justification for building the 'Leader' was that the Southern's Traffic Manager required a 'Passenger Tank Engine', as a replacement for ex-LSWR 'M7' 0-4-4Ts.

Bulleid had sketched out some ideas in 1944 for a steam locomotive capable of running in either direction, based on the 'Q1' 0-6-0 but carried on two power bogies. From the start, Bulleid seemed determined to use his remit to evolve an unconventional design far-removed in purpose from the tasks performed by 'M7s', as well as exploiting the changed requirements of the Traffic Department. Another influence was the Government's intention to support a scheme for the large-scale conversion of the steam fleet to burn oil and with which Bulleid was closely involved. Accordingly, a double-ended locomotive with a centrally-positioned firebox had become a practicable proposition.

By 1946, the project had materialised as a locomotive making full use of its total weight for adhesion and braking, capable of being driven in either direction, and mounted on two bogies, each separately driven from two, three-cylinder engines, with oil bath lubrication, and with chains in place of coupling rods. By using sleeve valves in place of piston valves, it would be easier to fit the engine into the space of the power bogie. Then it was decided that the design should be arranged for coal-burning.

With a cab at each end, at first sight the 'Leader' resembled a diesel or electric locomotive but the cabs were very different to those of a non-steam locomotive. All controls were duplicated in either cab and amidships was the fireman's position, half the width of the locomotive, to the rear of the firebox, and in front of the coal and water spaces. Ventilation was totally inadequate and the fireman's job was worse than that of a ship's stoker.

In late 1946, the ordering of material for the locomotives began, manufacture started in July 1947, and the first set of frames was laid at Brighton Works during May 1948. Trials began from Brighton in the summer of 1949, and No 36001, the first of what was being described as the 'Leader' class, was moved to Eastleigh in April 1950 for further testing.

By now, Bulleid had left the Southern Region for his stint with Córas Iompair Éireann, and responsibility for the 'Leader' passed to the Railway Executive. A preliminary report dated March 1950 set out the pros and cons of the project, and warned that much work and money would need to be expended to make the 'Leader' a practicable locomotive. In March 1951 came the inevitable decision to abandon the project.

From LMS to London Midland in the 1940s

The greater size of the LMS meant that the company faced a much harder task when recovering from the effects of the war. Then, during the process of nationalisation, there was the question of how LMS territory would be distributed under the new regime. First thoughts were that the English portions of the LMS would become a North Western area. There would be a Scottish area made up of LMS and LNER lines. Then came a proposal for the southern part of the LMS to be made one Region, another Region being formed from LMS territory in northern England,

Below:
Rebuilt 'Royal Scot': No 6152 *The King's Dragoon Guardsman* heads a Euston-Blackpool express away from Northchurch Tunnel, north of Berkhamsted, on 17 June 1947. This engine had been rebuilt with a taper boiler in August 1945. *E. R. Wethersett*

and perhaps with its headquarters in Manchester. There was no change of plan for a Scottish Region. The organisation eventually adopted was a North Eastern Region based on York, but all LMS lines south of the Border became the London Midland Region (LMR). Subsequent changes were made in 1949 when the former London Tilbury & Southend lines were transferred to the Eastern Region (ER) and the Central Wales line was ceded to the Western Region, while the LMR gained the system operated by the former Cheshire Lines Committee.

In its last couple of years the LMS was carrying more passengers than in the immediately prewar years but they were conveyed in fewer passenger trains. More freight traffic was being handled. Yet the resources at the Railway's command were rundown, and both the years 1946 and 1947 were marred by serious accidents on the Trent Valley main line. As in the case of the other railways, the LMS had appointed various committees during the war to examine postwar development. One had considered modular designs of stations, and one structure was erected at Queens Park, just south of Willesden Junction, and became notorious because it was never served by trains; another similar building was completed - and used - at Marsh Lane on the Liverpool-Southport line.

As the LMS gave way to British Railways, enthusiasts were probably most interested in changes affecting the locomotive fleet. New directions were being taken with both steam and diesel traction. Considerable effort was expended in ensuring that before the LMS ceased to exist it would gain kudos from having completed the first main line diesel locomotive for service in Britain. This was the 1,600hp diesel-electric Co-Co No 10000 which on 8 December 1947 was driven out of the Diesel Shop at Derby Locomotive Works by H. G. Ivatt, the CME of the LMS. The planning for No 10000 had begun in 1945. While the mechanical parts were designed by the LMS, the Company had worked closely with The English Electric Co Ltd which supplied the diesel prime mover and all electrical control and power equipment, as well as generally contributing to the design of the locomotive.

H. G. Ivatt was responsible for the introduction of

Below:
The fearsome winter of 1947 caused considerable difficulties for the railways as they struggled to recover from wartime wear and neglect. This is a scene from that year at Dent station on the Settle & Carlisle line, notable for being the highest main line station in England.
Science & Society Picture Library RH2/A300019

several new steam locomotive designs, as well as undertaking improvements to existing ones. Some of this work had begun at Derby during the war when the LMS took a fresh look at steam locomotive design with the principal aims of reducing both servicing and maintenance. This was when Charles Fairburn had taken over as CME in succession to Sir William Stanier. The design which was to be associated with Fairburn was a revision of the Stanier 2-6-4T. The main alterations comprised a shortened coupled wheelbase and reduced overall length, which effected savings in steel.

The concern to reduce the time spent on servicing engines was influenced by the appearance in Britain of the 'S160' 2-8-0s built by American locomotive builders for the USA Transportation Corps. They were built for wartime operation on the Continent with the opening of a Second Front. Designed for maximum accessibility and easy maintenance, these engines reflected the requirements of US railroads which specified self-cleaning smokeboxes, rocking grates and hopper ashpans to make it easier and less irksome at the end of their daily duties when 'disposing of locomotives on shed. After trials, such features were fitted to all new LMS engines built after 1945.

Another objective was to increase the mileage achieved by engines between their visits to main workshops. On older engines particularly, axlebox wear had largely dictated the intervals between shopping, and if engines could go 100,000 miles between main works overhaul this would allow appreciable financial savings to be made. Such thinking led to the use of manganese steel liners in axleboxes, and a greater use of roller bearing

axleboxes. Careful attention was also paid to the detailed design of many other components and the use of superior materials.

Improvements of this sort were progressively incorporated in new locomotives built from the middle war years, principally Stanier 2-8-0s and 4-6-0s. From 1947, some of the Class '5' 4-6-0s were turned out with various other special fittings - such as roller-bearing axleboxes, Caprotti valve gear, steel fireboxes and double chimneys and sometimes with a combination of these features. The idea was to study the advantages of incorporating these innovations in general locomotive production.

More general aspects of locomotive efficiency could be studied under controlled conditions at a purpose-built testing plant. In 1937, the LMS and LNER had agreed jointly to build one at Rugby but its construction was interrupted by the war and it was not completed until 1948. BR had already embarked upon the inter-Regional Locomotive Exchanges of April-September 1948 when 14 locomotive classes from all Big Four companies ran comparative trials on several routes. Nothing which emerged from the trials had a major influence on the introduction of the new BR Standard steam locomotive designs.

During the war, the LMS had started work on new locomotive designs and some strange proposals were briefly considered, including an 0-6-0 with a taper boiler and a modernised version of a Lancashire & Yorkshire 2-4-2T. These evolved respectively into two thoroughly practical Class '2' types - a 2-6-0 and a 2-6-2T generally referred to as Ivatt designs; the first examples of both appeared at the end of 1946. Closely following them was a Class '4' 2-6-0 notable for a high running-plate of US pattern and a double

blastpipe and chimney. The first few of this type went into service in the last month of the LMS. Another new locomotive design during the period was the Class 'WT' 2-6-4T built for the Northern Counties Committee, 18 of which were completed between 1946 and 1950. These had the same 6ft diameter driving wheels and parallel boilers as the 'W' class Moguls built from the mid-1930s.

Between 1940 and 1946, 16 Stanier 'Princess Coronation' Pacifics had been constructed, and the last seven (Nos 6249-55) had lacked the streamlined casing. The streamlined engines were 'defrocked' between 1946 and 1949. In late 1947/early 1948 had come another two (Nos 6256 and 46257), the design of these Pacifics having been modified with the provision of roller bearings, rocking grates and hopper ashpans, as a result of which the rear end of the main frames and cab were changed, and the engines had a new type of the trailing truck. No 6256 of this pair was named *Sir William Stanier* and both engines were intended to provide a comparison in service with the LMS main line diesels, the second of these, No 10001, entered traffic during 1948.

The war years had seen the reboiling of two LMS 4-6-0 designs. As mentioned in Chapter 3, the deficient steaming experienced with the '5XPs' or 'Jubilees' was a matter of great concern. A number of changes were made to the design of the boiler used and also the draughting of the engines, and the final series of 'Jubilees' completed in 1936 incorporated a number of modifications.

Towards the end of 1935, the high-pressure compound 4-6-0 formerly named *Fury* re-entered service, having been largely reconstructed, including the provision of a new type of taper boiler to a larger diameter than that fitted to the '5XPs'. Renumbered 6170 and named *British Legion*, as first modified this engine had a single chimney and was devoid of smoke deflectors.

Before long, the Development Drawing Office at Derby was working on schemes to fit a larger boiler on to the chassis of a two-cylinder 4-6-0. The idea was to dispense with the inside cylinder of the '5XP' class, to save weight, and thereby permit the use of a shortened version of the taper boiler used for No 6170. The gradual lifting of civil engineering restrictions meant that it was possible to fit this boiler to an unconverted three-cylinder '5XP'.

During 1939, approval was received for rebuilding two rebuilt 'Jubilees' with the new boiler fitted with a double blastpipe and double chimney. The rebuilding also involved other changes to the engines' structure. Yet how was a project to rebuild the two 'Jubilees' justified at the onset of World War 2? Locomotive boilers came due for renewal at intervals of 20 years or so, and there was an annual programme for new boiler construction, to provide renewals and spare boilers. The latter were provided to reduce the time spent by engines undergoing general overhaul in works.

The new boilers were completed in October 1941, the LMS having decided to convert 32 locomotives to take what were described as 'later types' of boiler, with the consequence that the same number of 'earlier type' boilers would be released as spares. That total of 32 included the first 'Jubilees' chosen for conversion - Nos 5735 *Comet* and 5736 *Phoenix* - which came into Crewe Works for Heavy General overhauls early in 1942, and returned to traffic in the May of that year. By 12 months or so, these reboilered 'Jubilees' preceded the entry into service of the rebuilt 'Royal Scots' which carried the same type of boiler. Their existing boilers had become due for replacement and so it made sense to use the new design rather than renew like with like. The conversion of the 'Scots' continued in the early postwar years, as did the reboilering of 18 'Patriots which were dealt with in 1946-49.

The entry into service of so many modern engines during 1945 and 1949 displaced many old favourites

which as a result of the rigours of wartime service had become sadly rundown. The Midland and LMS Compounds were relegated to less demanding duties but by 1949 a clean sweep had been made of the remaining ex-LNWR express engines. The final members of the 'Prince of Wales' class were withdrawn that year, so were the sole surviving examples of the 'Precursor' and 'Claughton' classes. The last two 'George V' 4-4-0s had been condemned the previous year and the last Whale 19in goods 4-6-0 just made it into 1950.

Of the older classes of Midland engines, the final three Johnson 2-4-0s were withdrawn by 1950, and all but one of the three surviving Kirtley double-framed 0-6-0s. The numbers of the older Midland Class '2' and '3' 4-4-0s were heavily reduced, as were the ex-LNWR Webb 2-4-2Ts and 0-6-2T 'Coal Tanks'. Many secondary classes of LNWR design were reduced to last survivors in the late 1940s, including the earlier examples of ex-LT&SR 4-4-2Ts, and a clutch of LNWR classes - the 'Coal Engine' 0-6-0s and 0-6-2Ts, and the 0-8-2Ts and 0-8-4Ts. The L&YR 0-8-0s were eliminated.

Of the Scottish pre-Grouping types, the more numerous classes such as the Caledonian Pickersgill 4-4-0s, various classes of 0-6-0s, and 0-6-0T and 0-4-4T

types had several years' life in prospect. Less numerous classes were ruthlessly extinguished, one reason being that improvements to track and bridges had allowed the Stanier Class '5s' to work over more routes. Classes extinguished by 1950 included the Highland 'Loch' 4-4-0s and 'Clan' 4-6-0s. The very last G&SWR engine in traffic - an 0-6-2T - was another casualty.

South of the Border, some ex-LNWR classes were kept hard at work, including the 'Cauliflower' 0-6-0s which worked passenger and goods turns over the attractive Cockermouth, Keswick & Penrith line and the Webb 2-4-2Ts employed on push-pull trains in the West Midlands. Replacements for these classes were the Ivatt 2-6-0s and 2-6-2Ts respectively. Well distributed among former LNWR sheds were the '6F' and '7F' 0-8-0s and for at least another decade they would continue to uphold the North Western banner.

The late 1940s saw the construction of a large number of standard LMS corridor and non-corridor coaches. Generally, these followed the Stanier-period designs with some tidying up of internal and external details and were distinguishable on account of their porthole-shaped windows for toilets, and at vestibule ends. These coaches had wooden body framing with steel panelling, but one batch of corridor composites built in 1949/50 was of all-steel construction.

By nationalisation, there was effectively a typical LMS train comprising a Stanier '5' 4-6-0 at the head of modern steel-panelled coaches. Efficient and comfortable, it might be found working variously from Inverness to Wick, Manchester to Barrow, St Pancras to Leicester, Shrewsbury to Swansea or Bath to Bournemouth. LMS rolling stock travelled even further than that because it also regularly reached the East Coast at Yarmouth Beach station, and travelled deep into the West Country with the Bradford to Paignton express.

From LNER to Eastern and North Eastern in the 1940s

The LNER had tried hard to arrive at the priorities for its operation in the postwar world. Some of its visions, of electrified lines leading from its terminal stations into the suburbs were not to be realised but some idea of its general approach for the future can be gained from the Liverpool Street-Shenfield electrification.

During 1937, a start had been made on the 1,500V dc overhead electrification scheme whose preliminary associated works included considerable changes to the way and works of the Colchester main line. Combined with resignalling and improvements to the station's facilities, the track layout at Stratford station was extensively altered to achieve several objectives: to eliminate as far as possible all conflicting train movements, to allow the speed limit through the station to be relaxed, and to accommodate the extended Central Line tube service.

Before the electrification scheme had been approved, the main line had been quadrupled as far as Shenfield, where there was now a dive-under for Southend branch trains. Romford and stations east to Shenfield had been reconstructed. As part of the electrification work, the inner suburban stations such as Maryland and Forest Gate were reconstructed in a bright new style which obviously aimed to set an LNER standard. One of the biggest jobs associated with the electrification scheme was the construction of a flyover taking the slow lines across the fast line

Below:
Near Berwick-upon-Tweed on the East Coast main line, Thompson 'A2/3' Pacific No 521 *Watling Street* heads an up express on 25 August 1947. This was one of the 'A2s' to be built new, as opposed to a reconstruction of the Gresley 2-8-2s or 2-6-2s. *E. R. Wethersett*

between Manor Park and Ilford. By late 1946, tube trains had begun calling at Stratford station. Ilford flyover was completed the following year.

Originally, the intention had been to convert existing steam-hauled, teak-bodied suburban compartment coaches for the new electric trains but, in 1939, and before placing the contract with the suppliers, Metro-Cammell and Birmingham RCW, the LNER had a change of plan. The 92 multiple-unit trains would now be completely new, of all-steel construction, with sliding passenger doors and large windows. Eight similar units would be built for suburban workings at the western end of the Manchester-Sheffield/Wath electrification. These contracts were suspended in 1940, and reactivated and revised in 1946. The Liverpool Street-Shenfield electrification was brought into use in September 1949.

It took longer to complete the Manchester-Sheffield-Wath electrification although before 1939 work had begun on installing the overhead masts and on some other tasks. Included in the route's modernisation was the construction of a new Woodhead Tunnel through the Pennines and total resignalling. By the time work had restarted after World War 2, some of the more optimistic forecasts for traffic increases were revised and the extension of electrification to Trafford Park and into Manchester Central was deferred, while the

number of new electric locomotives was cut from 27 to just seven. It was September 1954 before the electrification of the main line was complete.

After the war the LNER attempted to return prewar standards to the East Coast main line. But with the increased loadings of the principal services an early return of the handsome streamlined expresses could not be contemplated. Instead of the 'Silver Jubilee' there was now a Pullman express between Newcastle and King's Cross and return. This was the 'Tees-Tyne Pullman' which ran to similar times as the streamliner and made its first public journey on 27 September 1948. The prewar 'Yorkshire Pullman' had already been reintroduced on 4 November 1946, at first to Leeds and Harrogate, then later to Hull; the Pullman was however suspended during the coal crisis of mid-1947.

The wartime planning committee had recommended the adoption of all-steel construction but instead the LNER's postwar coaches had to make do with wooden body framing and steel panelling.

Below:
First in the formation of this King's Cross-Edinburgh express of 15 July 1946 seen near Peascliffe Tunnel, north of Grantham, is one of the LNER's new steel-panelled corridor coaches, but the others in the train are late Gresley vehicles. The engine is Gresley 'A3' Pacific No 2580 *Shotover*, it was renumbered 81 at the end of 1946 and later became BR No 60081. *E. R. Wethersett*

The internal layout of side-corridor coaches was planned so that passengers would never be more than one compartment from an entrance vestibule. The new stock was very different in appearance from the Gresley teak-bodied vehicles but, having decided not to depart from its traditional 'livery', there was no option but to paint the steel-panelled coaches with a grained and scumbled finish to represent varnished teak.

The prototype of this new design appeared early in 1945. Between then and 1947, production examples had entered East Coast service. At regular intervals, beginning in 1924, the LNER had renewed the sets of coaches used on the 'Flying Scotsman'. In 1948, there came the first complete sets of postwar coaches which featured the double-glazing and pressure ventilation, as indeed had the latest of the LNER's prewar stock. In summer, the prewar 'Flying Scotsman' had run nonstop between King's Cross and Edinburgh. From May 1949, the former 'Junior Scotsman' was retimed to leave earlier and, under the title of the 'Capitals Limited', this train was the one making the nonstop runs. It was also equipped with the double-glazed, pressure-ventilated postwar stock and, with a clean 'A4' Pacific at its head, made a fine advertisement for the ambitions of the postwar LNER.

The Gresley 'A3s' and 'A4s' remained well in evidence on the principal express trains, despite the introduction of the Thompson Pacifics. On Gresley's death in 1941, Thompson had taken over as CME of the LNER. From the beginning the new CME made it clear that he was determined to introduce new standard designs of engine.

Although the LNER had done well to renew its first-line locomotives during the 1930s, it was a different matter with the secondary types. There was a pressing need to replace the many different pre-Grouping 4-6-0 classes, the 200 Atlantics and the 23 different classes of 4-4-0 which collectively totalled over 500. There were also numerous obsolescent 0-6-0s. In wartime conditions the derived motion fitted to many of the Gresley engines was difficult to maintain satisfactorily and for his new designs Thompson listed simplicity of handling and maintenance as his priorities. The best features of existing designs would be perpetuated in preference to designing anew.

Thompson identified priorities for three new types of locomotive: a general utility engine for mixed traffic duties; a powerful engine for the principal passenger duties and another type for fast passenger and fitted freight work; and a light mixed traffic engine, selected from existing designs. Gresley had produced the 'V4' 2-6-2 class to meet the last of these needs but only

two were built. Thompson evidently thought that a simpler engine was required.

The general utility engine emerged as the two-cylinder 'B1' 4-6-0, the first being No 8301 *Springbok* which was ready by the end of 1942. To reduce design and manufacturing costs, the 'B1' utilised the boiler design fitted to the Gresley 'B17', but with pressure increased to 225lb, and the same wheel centres of the 'V2' so the 'B1s' shared the same 6ft 2in diameter driving wheels. Another cost saving came with using fabricated members in place of steel castings. The 'B1' looked neat and traditionally British in comparison with some other wartime designs, and illustrated the continuity in design from Great Northern days to the very different conditions of wartime.

The last of the initial batch of 10 'B1s' was turned out in 1944. These engines were distributed around the system to assess their capabilities, sometimes

Above:
The 'B1' 4-6-0s featured in the 1948 Locomotive Exchanges, and No 61292 seen here underwent trials between Perth and Inverness. On 14 July 1948, it is seen at Perth station shunting the through coaches from Glasgow Buchanan Street on to the stock of the 4pm Perth-Inverness. This was a familiarisation run as the dynamometer car was being used for another of the engines being tested over the route. *Gavin L. Wilson*

running trials against older classes. Beginning in 1944, quantity production of the 'B1s' commenced and by the time that the LNER had ceased to exist, there were over 270; construction finished in early 1952 when there were 409 'B1s' - one had been written off earlier after a bad accident. The 'B1s' quickly became recognised as competent on all sorts of duties. On the former Great Central main line they could be found at the front of the two named expresses - the 'Master Cutler' and the 'South Yorkshireman' - until train loads began to tax the 4-6-0s and they gave way to 'A3'

Pacifics transferred from elsewhere. In the early postwar years 'B1s' were the choice on the former Great Eastern main lines for the principal express duties, such as the 'East Anglian' and Liverpool Street-Parkeston Quay boat trains.

For the two powerful types of engine envisaged by Thompson, the first examples were produced by extensive rebuilding of two Gresley designs. The 'P2' 2-8-2s had entered traffic from 1934 and were employed on the Edinburgh-Aberdeen main line where in time their availability declined and their maintenance costs rose, not to mention some major and expensive failures. Thompson chose the six 'P2s' as subjects for rebuilding in 1943/44 as Class 'A2/2' mixed traffic Pacifics. Another rebuilding project produced Pacifics from the last four 'V2' 2-6-2s under construction. These retained the 'V2' boiler and firebox, now fitted with a new smokebox with double blastpipe and chimney; these became 'A2/1s' and entered traffic in 1944/45. The 'A2/2s' were put to work on their old stamping-ground while the 'A2/1s' were used between Newcastle and Edinburgh to provide data for the design of the standard 'A2/3' built new from 1946. The first of these was No 500 *Edward Thompson.*

The 'A2s' were truly mixed traffic engines and had 6ft 2in diameter driving wheels, but the other Thompson Pacific design was the 'A1' with 6ft 8in driving wheels primarily intended for express passenger work. The prototype 'A1' was the original Gresley Pacific, No 4470 *Great Northern.* It was extensively reconstructed so that little of *Great Northern* can have remained. New main frames were fitted, together with new cylinders with the leading bogie placed ahead of them, three sets of Walschaerts valve gear instead of Gresley conjugated motion, and the boiler was of the type fitted to 'A4s'. The finishing touch was a double blastpipe and double chimney. Although LNER official publications indicated that No 4470 was to be streamlined, this never materialised.

The production 'A1s' were not built until after Thompson's retirement and modifications were made to the design as compared with No 4470. The so-called Peppercorn 'A1' was much nearer to the sort of Pacific that might have been designed by Gresley and 49 were built during 1948/49. The 'A1s' proved to be particularly successful on heavy King's Cross-West Riding and overnight East Coast sleeping car expresses.

Thompson's plan for a range of standard LNER locomotives included adaptations of existing designs. Of the completely new designs there was the 'B1' and also the 'L1' 2-6-4T, a powerful locomotive which was intended both for use as a mixed traffic engine and to displace tender engines from outer suburban passenger workings. Although on trials the prototype performed well, in general service the 'L1s' were less satisfactory, with a tendency to slip badly on starting, and to be heavy on maintenance. Of the modified existing designs, one of Gresley's 'K4' 2-6-0s built for

Below:
The pioneer Thompson 'L1' 2-6-4T No 9000 (later BR No 67700), in LNER lined apple-green livery, and photographed while engaged in trials from Gateshead shed during July 1946.
E. R. Wethersett

service on the West Highland line was modified as the prototype two-cylinder 'K1' 2-6-0, 70 production engines being built after Thompson's retirement. The Robinson 'O4' 2-8-0 was used as a basis for the 'O1' standard engine to work coal and mineral trains. A new boiler and new cylinders were fitted. Robinson's 'J11' 0-6-0 was rebuilt with piston valves and a higher pitched boiler to become the 'J11/3'.

Some of Thompson's rebuilding projects were questionable, both as regards value for money and their practicality. That was particularly true of the single Great Central four-cylinder 'B3' 4-6-0 rebuilt with a 'B1' boiler and two cylinders, of the reconstruction of ex-Great Central 'Q4' 0-8-0s to produce the 'Q1' 0-8-0T heavy shunting engines, and of the rebuilding of three-cylinder 'B17' 4-6-0s as two-cylinder 'B2s'. As with some other engine rebuilding projects with which Thompson had been involved before 1939, there was more than a hint of 'new wine in old bottles'; in the long term, the rebuilt locomotives tended to suffer from cracked frames.

By 1947, there were about 400 of the new standard Thompson engines in service, whether of new construction or rebuilt. The intention was that over 1,000 engines of 11 existing LNER classes would be retained in service but reboilered when the existing boilers wore out. To that end, a range of standard boilers was selected for this purpose.

The changes to the passenger rolling stock were less drastic although the construction of the standard postwar corridor coaches enabled a start to be made in clearing out some of the oldest vehicles. The standard corridor coaches for general service lacked the pressure ventilation of the 'Scotsman' sets and, until 1949, what is sometimes called the Thompson or 'transverse corridor' stock had square-cornered windows but, to reduce corrosion of the steel body panelling, the windows of the later vehicles built up

until 1950 had radiused corners. There was nothing special about the neat and harmonious non-corridor postwar stock which put paid to many pre-Grouping vehicles and continued to be built until 1953. The teak-panelled coaches in special sets built for 'East Anglian' and 'Hook Continental' in prewar days returned to their original duties. The stock of the LNER's high-speed expresses was never again marshalled in complete sets but several vehicles from the former 'West Riding Limited' were among the first coaches on BR to carry the new crimson and cream livery and in 1949 they were allocated to the new 'West Riding' express.

With the formation of the Scottish Region (ScR), LMS and LNER lines came under the same management and some duplication of services and facilities was cut out and in one or two areas LNER locomotives and rolling stock gradually gave way to LMS stock. In general, services were unaltered on former LNER routes such as the West Highland Line, from Craigendoran to Fort William and onwards over the Mallaig Extension. In the late 1940s the West Highland line was worked with Gresley 'K4' 2-6-0s which had been built for this route's passenger trains and which shared duties with the 'K2' 2-6-0s (specially fitted with side-window cabs). Newly constructed 'B1' 4-6-0s and 'K1' 2-6-0s then took over workings on the line and the 'K4s' were relegated to freight turns, and increasingly to the Mallaig Extension.

The former Southern Area of the LNER had become the Eastern Region of British Railways, and the previous North Eastern Area now comprised the new North Eastern Region. In practice, the North Eastern Region was not entirely autonomous. Some functions were managed jointly with the Eastern Region, in particular mechanical and electrical engineering, and the Regions co-operated closely on the operation of the East Coast main line.

5. **The 1950s:**

From Britannia to Ark Royal - the Modernisation Plan Era

British Railways started the 1950s with a Labour Government in power but the decade concluded with the 'never had it so good' philosophy of Prime Minister Harold Macmillan, at the head of a Conservative Government that had become impatient with the dismal financial performance of the nationalised railways. In the meantime the public had moved decisively towards personalised transport once petrol had come off ration and hire purchase made it easier to buy a new car. There were 2.3 million private cars in 1950 and 5 million nine years later; 916,000 goods vehicles in 1950, but 1.3 million in 1959. The railways were facing threats from other forms of transport, too; even internal air services were cutting into Anglo-Scottish passenger traffic. Under the Labour Government's scheme of things the British Transport Commission (BTC) would have been the controlling force in inland transport and there were proposals for

integrated transport in selected areas. The Conservatives dismantled much of the centralised bureaucracy of transport and freed the road hauliers from restrictions. The railways were now exposed to competition from operators who were less hide-bound than they were.

The Government had acknowledged its heavy reliance on the railways during World War 2 and the backlog of investment. Relatively low levels of investment were made available in the first years of nationalisation, and the system continued to disinvest. Most of the major improvement schemes dated from before 1939. Besides, the 1945-51 Government faced

Above:
The East Side platforms of London's Liverpool Street station on a peak Saturday in 1952. Crowds of passengers patiently await their trains to Clacton-on-Sea and Southend. Sun streams through the windows fronting Bishopsgate, at a location that has greatly changed in recent years. *R. E. Vincent*

Above:
One of the first main line electrification projects to be completed by BR was the Manchester-Sheffield/Wath scheme. This is Mottram marshalling yard on Manchester's outskirts. Alongside comes a Sheffield-bound fast train headed by one of the Co-Co electric locomotives. *G. Richard Parkes*

more important considerations, such as the devaluation of sterling and rearming for the Korean War. After several years of deliberation, the BTC published its Modernisation Plan early in 1955. In its own words, this set out the basis of a 'thoroughly modern system' and envisaged that over the next 15 years £1,240 million would be spent to achieve this ideal. The Plan was criticised for being a shopping list of projects which BR was keen to complete but for which it lacked detailed costings. None the less, the Government supported the Plan but of course a commitment over such a long period was subject to changing political fortunes - and BR's own performance. In the summer following the launch of the Plan there was the damaging footplatemen's strike. Even if traffic was not directly lost, it taught the railways' customers that it made sense to have an alternative - just in case.

The Modernisation Plan dominated railway thinking through the late 1950s and into the early 1960s. If there were delays to trains the stock answer to enquirers was that modernisation work was in progress. Some schemes seemed to take ages to be completed. Changes were made to aspects of the Plan as the years passed but its basic strategy did not alter. Of the £1,240 million total investment (say, £18 billion in today's prices), the money was distributed reasonably equally among the various programmes.

In any case, spending was spread over 15 years and all six Regions. As originally presented, the lion's share of investment would have gone on the renewal of passenger and freight rolling stock, the next biggest area being electrification. Here priority was given to electrifying the LMR's Euston-Manchester/Liverpool lines, the ER's King's Cross-Leeds main line and Great Northern suburban lines, much of the former Great Eastern routes nearest London and the LT&SR system, all SR lines east of a line drawn from Reading to Portsmouth, and much of the Glasgow suburban network.

Dieselisation was listed only after rolling stock and electrification and the intention was that 2,500 main line diesels would be at work by completion of the Modernisation Plan in 1970. Even so, the Plan made provision for improvements at steam motive power depots, the idea being that steam would remain at work until the 1980s - when the BR Standard locomotives would have been fully depreciated. Not far behind in the proportion of spending was the construction of new marshalling yards and

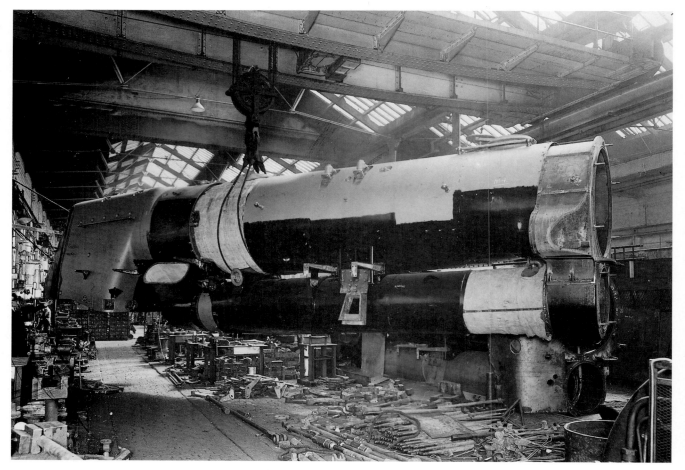

Above:
BR experimented with the use of the Italian Franco-Crosti boiler, in a bid to make the steam locomotive more efficient by using the waste gases from combustion to preheat water passing into the boiler. These boilers were fitted to '9F' 2-10-0s Nos 92020-9 but did not prove successful. *Ian Allan Library*

Below:
The all-steel BR Standard coaches were considerably safer than the pre-nationalisation designs. This is a corridor first under construction at Doncaster Works during 1951. *Ian Allan Library*

improvements to existing ones. Further down the list were the programmes for renewing BR's infrastructure, resignalling and the provision of new or rebuilt stations.

While much was made in the railway press of the Modernisation Plan's concern to replace steam traction, no commitment had been given to its elimination. Also, taken at face value the Plan was more concerned to improve freight handling and equipment than everything else. That was not surprising, seeing that merchandise traffic was fast being lost to road transport.

Many of the investment schemes of the Modernisation Plan were completed late and over-budget. By the time some were completed, the need had disappeared. Government called for reviews of the Plan when all the while BR's financial health was failing. The pressure was on after the disastrous freight results for 1958 when receipts were nearly £30 million down on the previous year, and the traffic carried was lower than in any postwar year. The Government was spurred to look further into the railways' finances, while the BTC called for the Regions to make economies and BR made plans to speed up its acquisition of main line diesels to replace steam.

That was the environment for BR in the 1950s - one in which many on the railways looked over their shoulders, yet much appeared unchanging to a casual observer. Whole swathes of the railway were steam-worked with locomotives and rolling stock that were obsolescent in 1939 - and whole sections of the network were pitifully under-used. Yet by the late 1950s there was much to hearten the enthusiast for steam railways. Many main line expresses were smarter than at any time since the late prewar years. Schedules had been speeded up, many top-link engines were beautifully turned out and performing better than ever before. Rolling stock was appearing in Regional colours. In late 1955 the BTC had informed the Regions that, subject to its approval, they could introduce special liveries for a limited number of locomotives and rolling stock employed on principal express passenger services.

This was the icing on the cake and it was only achieved with great effort on limited parts of the system. If there were complaints that engines were

Below:
BR made attempts to introduce more mechanical handling at goods depots, such as this fork-lift tractor being used to lift an experimental light-alloy container. *Ian Allan Library*

Opposite:
Members of a permanent way gang pose for a moment during engineering work near Hest Bank, between Lancaster and Carnforth on the West Coast main line. A 1950s view.
Science & Society Picture Library HLO 146

Opposite below:
In the days when railway enthusiasts thought nothing of swarming over tracks and platforms! The train is the 'Severn Venturer' of the 1950s, hauled by a Western Region '16xx' 0-6-0PT.
John Spencer Gilks

Left:
A Pullman Car Co steward serves American tourists with their drinks in the observation car of the Pullman 'Devon Belle' express on its way to the West Country in the 1950s.
Science & Society Picture Library 2168

Left:
During the early/mid-1950s, the London Midland Region enjoyed considerable success with its 'North Wales Land Cruise' trains for holidaymakers. The trains followed a circular route through the mountains and along the coastline and made use of cafeteria cars which had been converted from existing catering vehicles to provide self-service light refreshments, as seen here.
Science & Society Picture Library BTF 3238

Right:
One of the few formerly independent light railways to retain a public passenger service was the former Kent & East Sussex but at the beginning of 1954 this was withdrawn. Special passenger services and freight traffic survived longer on the section south of Tenterden Town. Seen in October 1953, this is High Halden Road station with its hand-operated signal, on the doomed stretch to Headcorn. *K. W. Wightman*

Below:
The railway preservation movement began in 1951 with the 2ft 3in gauge Talyllyn Railway. On 1 August 1953, the last train of the day to Abergynolwyn approaches its journey's end behind 0-4-0WT No 2 *Dolgoch*, built in 1866. *D. R. Forsyth*

not as clean as they might be, then the critics had obviously never seen the grime and squalor at motive power depots and generally looked 'behind the scenes' on the railway. Dick Hardy, that thoroughly professional railwaymen and objective commentator on his industry, provided a perspective on the BR of the time in his biography of Dr Richard Beeching: 'In the mid-1950s, before the extension of electrification in north-east London and the introduction of diesel traction, we had been in desperate straits. Short of men, materials, engines but never of courage and resourcefulness, we had moved from crisis to crisis.'

He was talking of the Great Eastern lines of the Eastern Region, arguably one of the most enterprising parts of BR of the time.

Of course changes were in hand, it was just that they were taking place relatively slowly, and in recent years the nationalised railway had lost sight of commercial and accounting disciplines. No one knew for sure how much it cost to move passengers and freight traffic. Most sections of railway were failing by some margin to cover their direct costs. Wage rates were not competitive with other industries. Despite its problems - one might almost say *because* of them -

this was the time of the growth of railway enthusiasm. The membership of railway clubs boomed. The platform ends of main stations were thronged by trainspotters, particularly on peak summer Saturdays and at favoured locations such as Paddington, King's Cross, Tamworth and Reading. The sales of railway books and magazines grew rapidly.

The Western Region in the 1950s

No doubt about it, the Western Region (WR) had some of the best-looking express trains of the late 1950s. They were a vast improvement on those at the start of the decade when, by official admission and as demonstrated in the Regions' league tables, timekeeping was poor. Western Region operating superintendents met during 1954 to talk of needing to pay special attention to the working of London and South Wales expresses following complaints of consistently late arrivals - on average these expresses were running 11min late.

The suggestion was that Western enginemen were not as keen as they might be. They were certainly conservative and sure of their prejudices, none more so when commenting on the design and performance of the BR Standard classes they encountered in their work. Some of the Standard engines allocated to Western sheds were seldom diagrammed for any demanding work, yet the Class '5' 4-6-0s sidelined by the WR were turning in excellent work on much more heavily graded lines on the Southern Region. One of *Trains Illustrated's* correspondents in 1955 described a WR driver whose style on any main line

locomotive, to quote, 'never varies. He crawls away from Newton Abbot as if he had all day ... and makes no attempt to rush Dainton (bank). By Stoneycombe his engine is usually blowing-off strongly; but will he use this steam by opening his regulator and advancing his cut-off? Not he! He sits on his seat smiling at nothing in particular, while the engine losing speed every minute, wanders round the sharp "S" bend and arrives at the tunnel (still blowing-off) at a walking pace.' Similar drivers could be found at the London end of the Region. While in some WR quarters there was bitter criticism of the BR Standard 'Britannia' Pacifics, crews at Cardiff Canton obtained some excellent work from them when working London expresses.

At Swindon Works, inspiration and careful experimentation were being harnessed to extract the best from steam locomotive designs in the face of a declining quality of coal. During the early 1950s, the Western Region Chief Mechanical Engineer's experimental and research section at Swindon led by S. O. Ell carried out research which aimed to increase engines' steaming rates, in particular the 'Kings' and 'Castles'. The stationary plant for locomotive testing

Below:
A good example of the smart Western Region named expresses of the late 1950s. This is the 'Cornishman', from Wolverhampton Low Level to Penzance, approaching Mangotsfield on the Midland Railway's West of England main line on 3 May 1958. 'Castle' No 5045 *Earl of Dudley* is at the head of BR chocolate and cream liveried coaches, with the exception of the leading GWR-design Hawksworth vehicle. *Ivo Peters*

Above:
A controlled road test being carried out by the CM&EE's staff based at Swindon Works. The engine is 'King' 4-6-0 No 6001 *King Edward VII* on the Bristol main line at the head of a 25-vehicle, 800-ton test train in July 1953. Instrumentation was placed behind the indicator shelter at the front of the engine. *Ian Allan Library*

Below:
A more typical WR express of the early/mid-1950s: this is a Paddington-Oxford-Worcester train, making its way down to Reading and photographed near Slough on 26 May 1953. The engine is Worcester shed's 'Castle' No 5086 *Viscount Horne* which had been reconstructed from 'Star' No 4066 in 1937; it was to be one of the first 'Castles' to be withdrawn from service by BR. At the front of the train of ex-GWR stock is a 'Toplight' corridor composite which would soon be withdrawn with the arrival on the WR of large numbers of BR Standard coaches. *E. R. Wethersett*

was on hand in the Works. The immediate conclusion of this research was that carefully devised modifications to the draughting of engines could significantly increase steaming rates and an improved performance would allow faster point-to-point timings to be introduced on WR expresses; it meant also that if saddled with inferior coal, engines would be better able to maintain most express schedules, if not the fastest.

The modifications were put on trial during controlled road tests out on the main line, and on the Swindon testing plant. By late 1953, enough 'Kings' and 'Castles' had been modified with improved draughting to allow a number of WR express trains to be accelerated as from the summer 1954 timetable, including the 'Bristolian' and the 'Pembroke Coast Express'. Further controlled road tests gave the

operating department confidence to reintroduce the 4hr Paddington-Plymouth timing of the down 'Cornish Riviera Express'. Such tests drew attention however to the scope for increasing the efficiency of draughting still further, and to allow the engine to be freer running.

This led directly to Swindon evolving a double blastpipe and chimney arrangement for both 'Kings' and 'Castles'. From 1955, first the 'Kings' and then the 'Castles' were modified in this way and members of both classes put in some very fine work before diesels began to replace them after late 1958 on the principal express duties. Several of these engines were timed at 100mph. Despite the undoubted capability of these double-chimneyed WR express engines, Regional punctuality continued to be below standard and, during a review in 1957, the WR

Right:
A melancholy row of condemned engines at Swindon Works 'dump' in May 1951. Leading is 'Saint' No 2936 *Cefntilla Court* and behind it are three 'Stars', tank engines and a 'Bulldog'.
R. H. G. Simpson

Below:
Having received a heavy overhaul during which it was fitted with a double blastpipe and chimney, 'Castle' No 5033 *Broughton Castle* stands outside Swindon Works on 16 October 1960. It lasted no later than April 1963.

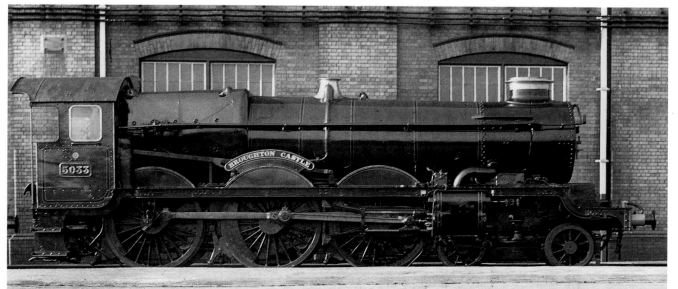

officers attributed this to mechanical defects on locomotives, poor quality coal, and frequent signal delays caused by heavy freight traffic. By then, the WR had been committed to a building programme for over 100 main line diesel locomotives. The names of the first few were challenging: *Active, Ark Royal, Bulldog, Conquest* and *Cossack*....

The research carried out at Swindon into locomotive draughting was applied to BR Standard classes such as the '4' 4-6-0s and '9F' 2-10-0s, which accordingly were fitted with double blastpipes and double chimneys. By the late 1950s the WR had received a number of '9Fs' in replacement of the earliest GWR '28xx' 2-8-0s, and the good performance of the new arrivals went some way towards winning WR enginemen's enthusiasm for the Standard classes. Not all Western drivers were prejudiced against other Regions' locomotives. Sometimes the volume of inter-Regional workings is overlooked. Bulleid Pacifics could be found at work on stopping trains between Exeter and Plymouth in the interests of maintaining knowledge of different types of engine in a situation when the South Devon main line might be unavailable. In the hands of Southern crews, 'Lord Nelsons' and, in later years, Bulleid Pacifics worked to Oxford with the through expresses from Bournemouth to Birkenhead, and Bournemouth to York. WR engines worked on to the Great Central main line with the Bournemouth-York train, and also the night-time Swindon-York passenger and parcels trains. Another line featuring through workings by 'foreign' engines was the North & West route southwards from Shrewsbury to Hereford and Newport. LMS 'Royal Scots' and 'Jubilees' appeared as far south as Pontypool Road on through trains from Manchester, Liverpool and Crewe and the West Country. During the 1950s the through working of engines with freight trains became even more common and a variety of unusual types travelled to South Wales in particular.

The Western Region was ever helpful with enthusiast societies seeking to run rail tours, and often accommodated requests for strange motive power. In the early 1950s, ex-GWR railcars were chartered for various excursions, there were tours of goods-only branch lines, and 'last appearances' by engine types shortly to be withdrawn. One institution began in the mid-1950s. This was the running of an excursion from Paddington to Towyn (nowadays spelt Tywyn) to take members of the Talyllyn Railway Preservation Society to the Society's AGM. In September 1955, 'Star' No 4061 *Glastonbury Abbey* had worked the TRPS special from Paddington to Shrewsbury where ex-LSWR 'T9' 4-4-0 No 30304 took over; it was assisted from Welshpool by a WR '90xx' 4-4-0. The following

Below:
Enthusiasts at work! The special train run annually during the 1950s to take members of the Talyllyn Railway Preservation Society to the AGM at Tywyn. This was the September 1955 train on its outward journey and here at Welshpool '90xx' 4-4-0 No 9027 prepares to come on to pilot ex-LSWR 'T9' 4-4-0 No 30304 to Tywyn. *T. E. Williams*

year's special once again made use of a 'Star' as far as Shrewsbury where an odd couple backed on to work to Towyn - ex-SE&CR 'D' 4-4-0 No 31075 and a Dean Goods. In 1957, the TRPS was double-headed west of Shrewsbury by an ex-Lancashire & Yorkshire 2-4-2T and a '90xx' 4-4-0, and for 1958 the special made a fine sight leaving Paddington behind one of the few remaining LMS Compounds.

The Western Region seemed to take railway enthusiasm to heart. Although the narrow gauge Vale of Rheidol line had reopened for summer tourists as early as July 1945, for years little attempt was made to promote it. For the 1955 season that stance changed, and extra trains were put on, and for the next season the coaching stock had been repainted in Regional chocolate and cream livery, and the line's three engines named; for 1957 they were given lined-out Brunswick green paintwork. These initiatives contributed to a slow but steady growth in the line's traffic during the next few years.

The Southern Region in the 1950s

Before World War 2, the Southern Railway did not reveal its strategy for progressive electrification of its network. Perhaps there was no such thing. Schemes for particular lines were submitted for approval by the Board of Directors, and either authorised or referred back. The 1946 plan which proposed the almost total electrification of all lines east of Salisbury was mentioned in Chapter 4. After nationalisation, however, all thoughts of electrification schemes were strictly under wraps.

The Modernisation Plan of 1955 promised third-rail electrification of the lines to the Kent Coast, onwards from the existing limits of electric working at Sevenoaks, Maidstone and Gillingham. From Ashford (Kent), electric trains were to work via Rye to Hastings. In the event, the Kent Coast electrification was split into two schemes: Stage 1 comprising the former London, Chatham & Dover main lines; Stage 2, the South Eastern main line to Folkestone, Dover and up to Margate. The Ashford-Ore electrification was later dropped.

Stage 1 saw electric multiple-units take over almost entirely from steam traction on passenger workings from June 1959 with electric and diesel locomotives used on freight duties. Then came Stage 2. For the moment, no other electrification schemes were discussed. Instead, the SR pressed ahead with introducing diesel traction elsewhere. Diesel-electric multiple-units were used to replace the 'Schools' class on the Charing Cross/Cannon Street-Tunbridge Wells-Hastings service in 1957/8. This represented a change of plan in 1955: the original intention had been to build new loco-hauled stock and continue with steam working. In Hampshire, diesel-electric multiple-units were used for local services radiating from Southampton; this scheme came in during 1957.

Below;
The 'Golden Arrow' is worked by a rebuilt 'Merchant Navy' 4-6-2 No 35015 *Rotterdam Lloyd*, climbing past Sydenham Hill with the down train early in 1959. *R. C. Riley*

Above:
At one stage, BR intended to re-equip the Charing Cross-Tunbridge Wells-Hastings service with new coaches, but to continue with steam haulage. Policy then changed to the provision of new diesel trains. On 12 July 1952, an up Hastings express climbs to Weald signalbox, between Tonbridge and Sevenoaks, behind 'Schools' 4-4-0 No 30905 *Tonbridge. E. R. Wethersett*

It was unclear if the Southern would adopt either diesel or electric traction on its South Western Division main lines where express services were largely worked by the still youthful Bulleid Pacifics. The Region had put into service three main line diesel locomotives, the last of them in 1954. Bulleid, proponent of steam though he was, had begun negotiations for the supply of the first two locomotives as early as 1947. The three SR diesels underwent trials on the Region and, when the Modernisation Plan was published, BR announced that it was intended to change over entirely to diesel traction in two areas, one comprising the main lines from Waterloo to Weymouth and to Exeter.

Perhaps it was relevant that during 1955 it was agreed to proceed with the rebuilding on more conventional lines of 30 Bulleid Pacifics, with equal numbers of 'Merchant Navies' and Light Pacifics. Altogether four authorisations were given for rebuilding all the 'Merchant Navies' and 60 of the Light Pacifics, at a total cost of £737,000 (say, £9 million in today's prices). It was not until the Bournemouth electrification scheme was being discussed during the 1960s that there were thoughts of eliminating steam traction on the South Western Division.

The Bulleid Pacifics, in original and now in rebuilt form, were effectively unchallenged during the rest of the 1950s but earlier the SR had operated both its own main line diesels and the pair of LMS-design examples on Pacific-worked express passenger turns on its South Eastern and South Western Divisions; it was stressed that these were trials only. Other interlopers included a pair of 'Britannias' - Nos 70004 *William Shakespeare* and 70014 *Iron Duke* - which were allocated to Stewarts Lane shed to work the

'Golden Arrow' and other boat trains. These engines remained on the Southern Region until 1958.

What of the fortunes of the other SR steam locomotives? In May 1949, one feature of the Regional allocation had been the number of engines in store. At main sheds and at Eastleigh running shed and sidings there remained not only the unwanted engines converted to oil-burning and now regarded as expendable but many other venerable types. A quantity of older engines, stored and in traffic, were however cleared out during the early 1950s when new engines of LMS design were delivered to the Southern Region. These comprised 30 Ivatt 2-6-2Ts built at Crewe Works, Nos 41290-41319, and 41 Fairburn 2-6-4Ts built at the SR's own Brighton Works, Nos 42066-42106. These engines went to Eastern and Central Division sheds but before long half a dozen or so of the 2-6-4Ts were transferred to the North Eastern Region. The others made a marked difference to the working of the semi-suburban Oxted lines and local trains in Kent.

One feature of Southern Region steam operations of the 1950s was a continued if slowly decreasing reliance on pre-Grouping classes of engine. On the Eastern Section, ex-SE&CR Wainwright 'C' 0-6-0s worked main line and branch goods trains, 'H' 0-4-4Ts could be found at work on push-pull trains (they were also used on the Central Division) and the rebuilt 'D1s' and 'E1s' as well as the 'L' class 4-4-0s were fully employed on secondary and some semi-fast trains, notably on the Chatham main line. 'R1' 0-6-0Ts continued to pound up the Folkestone Harbour branch with boat trains.

On the Central Section, although many ex-Brighton classes had been withdrawn, 'C2x' 0-6-0s could be found on many of the pick-up and branch goods trains, 'E4' 0-6-2Ts were hard at work shunting and on branch passenger and local goods work, and the powerful 'E6' and 'E6x' 0-6-2Ts were employed on the Deptford Wharf goods branch and on coal trains to Waddon Marsh power station. The 'K' 2-6-0s could be found working a variety of main line freight trains, and in summertime appeared on through holiday trains for part of their journeys from the Midlands to the South Coast.

Then of course there were the evergreen Stroudley 'Terriers'. 10 remained in capital stock through the 1950s and were used for shunting and local goods work from St Leonards and Brighton sheds, those at the last-named being employed on shunting at Newhaven. The majority were allocated to Fratton depot on the Western Section where they were used to haul all services on the Havant-Hayling Island branch. Three 'Terriers' were used for shunting at SR Works; few of us will forget the sight of DS377 in Stroudley-style livery at Brighton Works.

Larger Brighton passenger engines had become thin on the ground by the early 1950s. There were of course the Baltic Tanks that had been rebuilt by the SR as 'N15x' 4-6-0s. But top of the list were the Marsh Atlantics. Until the arrival of Fairburn 2-6-4Ts early in 1951, two Atlantics were used for Oxted line business

Above:
Fairburn LMS-design 2-6-4Ts were built by the SR's Brighton Works for service on the Region. No 42097 was photographed near Selsdon station on 7 July 1951, with an up Oxted line train formed of an ex-SE&CR 'Birdcage' set.
E. R. Wethersett

Right:
Until its closure in 1963, the diminutive ex-LBSCR 'Terrier' 0-6-0Ts worked the Havant-Hayling Island branch. No 32661 nears Havant from Hayling Island on 24 September 1959.
E. R. Wethersett

trains. Later, two were held as spares for working Newhaven boat trains, and the three others were at Brighton shed. An Atlantic was often provided to work the lightly loaded Brighton-Bournemouth through train and on summer Saturdays at least one would be turned out to work the through Midlands-South Coast holiday trains south of Willesden Junction. No 32424 *Beachy Head* became the last survivor and took its last bow in April 1958.

On former South Western territory, one or two Drummond classes, such as the 'M7' 0-4-4Ts and '700' 0-6-0s, were largely or wholly intact. The 'M7s' were at work on push-pull turns, doughtily working empty coaching stock into and out of Waterloo, assisting at the rear of trains from Poole up Parkstone bank, and bustling along with local trains between Exeter Central and Exmouth, and on other East Devon branches. Class 'O2' 0-4-4Ts were rarer, apart of course in the Isle of Wight where they almost monopolised train working. Other than west of Exeter, the 'T9' 4-4-0s had lost most of their regular passenger work with the advent of the Hampshire DEMUs, but they turned up at all sorts of places, whether on parcels trains into Portsmouth, or lending a touch of distinction at Reading South at the head of a train to Redhill. On this route 'T9s' were very much the exception for the various classes of Maunsell Mogul were the usual motive power. They worked the stopping passenger trains, freight turns and headed the through Birkenhead-Hastings/Kent Coast train on which they frequently turned in sterling performances between Reading and Redhill with loads of up to 11 coaches.

Enthusiasts were particularly fond of that line alongside the North Downs, and were also quick to glance out at Guildford where the steam shed was shoe-horned into a tight corner. There was a shed pilot here, formerly a shunter from the Southampton Dock Co but, for much of the 1950s, ex-LSWR 'B4' 0-4-0T No 30086. Guildford had an allocation of 'M7' 0-4-4Ts for the service to Horsham and other workings but ex-LBSCR 'E4' 0-6-2Ts were also used.

No new non-corridor stock had been built by the Southern Railway although it had rebuilt pre-Grouping vehicles in the 1930s for use on local and push-pull services. The push-pull sets were often a delight, usually made up of matching vehicles from one of the pre-Grouping companies but a few sets were a real mixture.

Some of the SR branches in the Home Counties were decidedly rural but the straggling lines west of Exeter, into North Devon, and away from Okehampton to Cornwall's North Atlantic coast seemed to be in the back of beyond. Padstow was 259¾ miles from Waterloo and as much as 3hr 20min travelling from Exeter Central. Not that this bothered the enthusiast bound for the far west to glimpse the three 80-year old Beattie 2-4-0WTs whose primary duty was hauling china clay traffic over the Bodmin-Wenford Bridge branch. Perhaps

Below:
Guildford, with its fondly remembered engine shed. WR '43xx' 2-6-0 No 6312 enters the station area on 15 May 1954 with a train from Redhill. WR engines had a regular Reading-Redhill return passenger working during the 1950s. *Stanley Creer*

The Southern Region's 'Withered Arm':

Above:
'N' 2-6-0 No 31833 approaches St Kew Highway station with a down freight train for Wadebridge. *B. A. Butt*

Below
A late 1950s picture of 'T9' 4-4-0 No 30338 with the 3.48pm train from Exeter Central, on arrival at Okehampton. *J. Davenport*

though enthusiast ambitions might be more restrained, to drop off at Axminster and enjoy an idyllic amble along the branch to Lyme Regis with an Adams 4-4-2T at the head of a single ex-LSWR coach. Happy days!

Some Southern railwaymen were far from happy about the change of management of former SR lines west of Exeter. After nationalisation these had passed to the Western Region but these so-called 'penetrating lines' remained under Southern Region control when it came to operating, and were worked with SR motive power and rolling stock. The lines reverted entirely to the SR in 1958, if no more than temporarily. The Somerset & Dorset was another 'penetrating line' whose northern section came under Western Region control, although operated by the Southern Region with a mixture of LMS and SR motive power and rolling stock.

The London Midland Region in the 1950s

Some 350 new engines of LMS design were built from 1950 but remarkably few of these came to the LMR. Those that did included 40 Ivatt '2' 2-6-2Ts, a dozen or so Fairburn 2-6-4Ts, 10 or so Ivatt '4' 2-6-0s, 28 Class '5' 4-6-0s including the final pair with Caprotti valve gear, 40 or so Ivatt '2' 2-6-0s and a handful of small 0-4-0STs. Over 200 engines from four of these classes were constructed at various BR workshops for Regions other than the LMR. At the same time diesel multiple-units displaced steam-worked local trains from a variety of LMR lines, beginning with the West Cumberland scheme of late 1954.

On heavily trafficked routes steam was as yet unchallenged. Take the southern end of the Midland main line. The express passenger trains from St Pancras-Sheffield, Manchester, Bradford and over the Settle & Carlisle line into Scotland were in the hands of 'Jubilee' and Class '5' 4-6-0s south of Leeds. Outer suburban trains from St Pancras-Bedford and semi-fast trains elsewhere were in the hands of '5s' or LMS Compounds. Many other local trains might

produce an ex-Midland or LMS '2P' 4-4-0, an Ivatt '4' 2-6-0, or one of the LMS 2-6-4T types. Passenger trains on secondary lines such as the Kettering-Cambridge service were worked by Ivatt '2' 2-6-0s. Branch lines from Bedford to Hitchin and Northampton and Wellingborough to Higham Ferrers were now in the hands of Ivatt or BR '2' 2-6-2Ts. The Midland 0-4-4Ts had been transferred away.

Train working on the 75-mile four-track Midland main line out of London to Glendon South Junction, north of Kettering, was dominated by the slow drags of southbound loaded coal trains and empty wagons going back to the collieries. The motive power varied, from Beyer-Garratts and Stanier '8Fs' to '4F' 0-6-0s. Workings were normally in two stages, from Toton, near Nottingham, to Wellingborough, and from Wellingborough to Brent, just north of Cricklewood. Fitted and partially fitted freight trains would produce '5' 4-6-0s and 'Crab' 2-6-0s.

South of Bedford the freight trains had to fit in among the outer suburban trains, the local passenger trains being more intensive south of Luton. Running

to St Pancras at off-peak times, and supplemented in weekday peak hours by St Albans-Moorgate trains, these inner suburban trains were worked mainly by Fairburn and Fowler 2-6-4Ts but the Moorgate trains were restricted to the Fowler 2-6-2Ts with condensing gear. The picture was rounded off by the appearance of Midland and LMS '3F' 0-6-0Ts on shunting and trip workings, by Midland '3' 0-6-0s on lighter goods trains and the occasional '2F' 0-6-0 numbered in the 58xxx series.

The picture remained much the same until dieselisation took hold from the end of the 1950s. By then, more BR Standard designs were at work on the Midland main line. The Class '5s' were to be seen on the less important express trains and semi-fasts, including some Caprotti valve gear engines allocated to Leicester, while '4' 4-6-0s were working from Bedford on St Pancras semi-fasts. The arrival of the Standard 4-6-0s had put paid to the Compounds which had become rare by 1958. The BR 2-6-2Ts were to be seen on the Bedford and Higham Ferrers branches. The biggest change had come to the Toton-Wellingborough-Brent coal trains for the Garratts had gone to the scrapyard, replaced by the '9F' 2-10-0s.

Cricklewood was a busy if hardly attractive place in the 1950s. There were the carriage sidings, the Express Dairy milk depot, next to which was the forgotten-looking station, the motive power depot and acres of sidings extending northwards to the North Circular Road. The railway was busy and constantly moving. It was a mistake to think that action was confined to the main line however, because there was the line from Brent Junction which crossed the LT lines near Neasden to end at Acton Wells Junction. It was and still is known as the Dudding Hill Loop. Acton Wells Junction gave access to Western Region lines, then another route was offered via Kew Junctions to Brentford and the huge marshalling yard at Feltham. This was one way in which freight traffic from the North and Midlands skirted London to reach the Southern Region and it was used also by excursion trains and troop specials.

To London railwaymen, the transfer freights loomed large in their operations. The cross-London

flow included a myriad of other freight workings, fish, milk and parcels trains for which there were strict instructions as to their operation. One of the longest cross-London workings was from Temple Mills, near Leyton, to Feltham, while other regular flows included Ferme Park (near Hornsey)-Acton, Ferme Park-Hither Green, Willesden-Feltham and Neasden-Feltham.

The many transfer freights from the Ferme Park Yards at the London end of the Great Northern main line passed by a different route across the capital, travelling by way of King's Cross York Road station, the Metropolitan Widened Lines, Farringdon, partly in tunnel to Blackfriars, and from there to yards at Battersea, Hither Green or various goods depots. Weight limits on Southern Region bridges meant that motive power was restricted to various ex-GNR or LNER tank engines but probably the 'J50' 0-6-0Ts were the favourites. The main loaded traffic was southbound, taking coal to domestic and some industrial consumers in the south and south-east. The rest was general merchandise.

The cross-London transfer freights were busy into the mid-1960s. These were mainly trains between marshalling yards: there was no other way to do it when the railway was handling thousands of individual wagon-loads. This was why the BR Modernisation Plan of 1955 put such an emphasis on improving facilities at marshalling and goods yards. A marshalling yard was somewhere to exchange wagons between trains, in order to form economic lengths of train. The Modernisation Plan aimed to replace the many individual yards at major centres by larger concentration yards. The Plan held out the promise of closing 150, to replace them with 55 new or reconstructed yards.

By the late 1950s the wagons using these modernised yards were slightly more up to date. From half a million in 1948, the number of wooden-bodied, ex-private owner wagons had been reduced in number to 364,000 in 1951 and less than 200,000 six years later. All these lacked through braking. A train of unfitted or largely unfitted wagons had to come to a stand before descending a steep gradient so that the guard could apply the hand-brake on all unfitted wagons, then return to his brake van to screw on his guard's van brake. Then the train would creep slowly downhill. There had to be a better way of operating trains in the 1950s! Private owner wagons remained for much longer in the shape of tank wagons used for petroleum products, tar and other chemicals, and other wagon types used for lime, limestone, salt and chemicals.

By the early 1950s many of the marginal branch lines were being closed down but the future of most of the network seemed assured. On some sections of railway, though, what was happening was perhaps a process of closure by degrees; some said, by stealth.

In the 1950 *Bradshaw*, Table 263 listed the service from Birmingham to Ashchurch via Redditch, Alcester and Evesham. Table 264 showed another way of reaching Ashchurch, via Worcester, Great Malvern and Tewkesbury. The line from Ashchurch to Tewkesbury had been opened by the Birmingham & Gloucester Railway as early as 1840, and in 1864 it was extended from a new station at Tewkesbury to Great Malvern by another company (later absorbed into the Midland Railway).

Right:
Rural Midlands branch line - Ivatt '4' 2-6-0 No 43017 brakes for the stop at Salford Priors with the 8.27am Birmingham New Street-Ashchurch, in readiness to collect a solitary passenger.
R. J. Rowley

On the east side of the Birmingham-Cheltenham main line at Ashchurch was the line via Evesham to Redditch. It was linked to the Tewkesbury and Great Malvern line by a spur that made a flat crossing with the main line. From Barnt Green, further north on the Birmingham-Cheltenham route, the line to Redditch was opened in 1859. The Evesham & Redditch Railway was completed throughout in 1868, and the final connection, between Ashchurch and Evesham, was the Midland Railway's contribution and it opened in 1864. The Redditch-Ashchurch line accordingly formed a loop line to the east of the main line and saw some use as a diversionary route by goods trains that were able to miss the Lickey Incline.

This is how the cut-back proceeded. No doubt in response to falling traffic, the five trains each way between Birmingham-Evesham and Ashchurch in 1950 had become four by 1959, with an additional Saturdays only working each way. The line from Great Malvern to Upton upon Severn was the first to go, all services being withdrawn in September 1952. Having been reduced to no more than a couple of trains each way, the Upton to Ashchurch passenger trains were withdrawn in 1961, and the goods trains were withdrawn by late 1964. The Ashchurch-Redditch line lost its passenger services in 1963 (buses had substituted for trains north of Evesham for a few months) and freight services were withdrawn south of Redditch in 1963/64. All that remains today is a stub of the line, to Redditch; at least it has been electrified.

After nationalisation, the Birmingham-Bristol main line remained under London Midland Region control although it ran through Western Region 'territory'. There had been attempts to transfer it to Paddington's control and finally, from February 1958, the WR took control south of Barnt Green. Some stations were repainted in WR brown and cream. The former Midland motive power depots at Bristol Barrow Road, Gloucester and Bromsgrove came under WR control. As if the change was being foreshadowed, from the previous year WR '94xx' 0-6-0PTs had been transferred to Bromsgrove to serve as banking engines on the 1 in 37 gradient on the Lickey Incline. Nowadays the Lickey is hardly noticeable from a High Speed Train but in steam days it was very different.

From 1920 until May 1956 the Lickey Banker was the Midland Railway's unique 0-10-0 which became BR No 58100 and was nicknamed *Big Bertha* or *Big Emma*. Complete with headlight (which never seemed to be used in later days) the 0-10-0 completed 838,000 miles before withdrawal, having spent almost all its time on the Lickey. Before it was built, standard Midland 0-6-0Ts were used as bankers and indeed their LMS successors continued at work until the WR pannier tanks came along in the late 1950s. *Big Bertha* was considered the equal of two 0-6-0Ts and its replacement was BR '9F' 2-10-0 No 92079.

Although other engines had been tried as Lickey bankers, including ex-LNER Garratt No 69999 (at intervals from the late 1940s until 1955, latterly as an oil-burner), none had stayed. Very few trains were allowed to tackle the Lickey unassisted although trials were conducted to see if this might be possible. With a train of 11 coaches the banking assistance might amount to three 0-6-0Ts, or *Big Bertha* and a single 0-6-0T.

The Lickey was of course a wonderful vantage point. Although always popular with railway

enthusiasts, its fame spread nationwide when the late Pat Whitehouse and John Adams featured it in their BBC television series *Railway Roundabout*. By the late 1950s, apart from the customary ex-Midland and LMS motive power on through trains, some strangers were creeping in on through workings, such as Eastern Region 'B1' 4-6-0s which were later to become regular visitors on summer Saturday holiday trains to and from the West which they worked to Bristol or Weston-super-Mare. Ex-GWR engines were not uncommon on special workings.

Some locations on main lines were virtually grandstands and in those days the regular cutting back or burning of lineside vegetation made it easier to watch trains. Then there were the major stations such as Carlisle Citadel which were highly regarded on account of the changes of locomotives on many workings. There was always the excitement of linesiding at locations where engines were working hard on adverse gradients, such as at Tebay where if not rushing the 1 in 75 climb to Shap Summit, trains would stop to gain assistance. Generally this was provided by a 2-6-4T at the rear although sometimes

Below:
Enthusiasts' vantage point at the cutting on the approach to Shap Summit as a Manchester/Liverpool-Glasgow/Edinburgh express storms past on 5 September 1957 behind 'Jubilee' 4-6-0 No 45635 *Tobago. Peter Groom*

Above:
Headed by a 'Princess Royal' Pacific, an express from Liverpool
Lime Street has arrived at Euston station's Platform 2 in July 1953
and now passengers move towards the platform end. Note the
antique bath-chair and numerous railway staff on hand!
Ian Allan Library

Below:
Continuing the Euston theme, an August 1955 view of the railway
at Bourne End, south of Berkhamsted. On the left, 'Princess
Royal' No 46210 *Lady Patricia* is speeding north with the 12 noon
Euston-Whitehaven/Morecambe while approaching is an up
express worked by a 'Jubilee' 4-6-0. *R. M. Newland*

Above:
One of the heaviest regular daytime express trains on the Western Division main line was the 'Merseyside Express', here speeding past Hillmorton, south of Rugby behind 'Princess Royal' No 46208 *Princess Helena Victoria* on 3 August 1957.
John P. Wilson/Rail Archive Stephenson

the tank came on as pilot to the train engine. Nowadays it is easy to forget the variety of motive power on our main lines. A day beside the main line at Tebay, say, could provide the sight of LMS, 'Britannia' and 'Clan' Pacifics, 'Royal Scots', rebuilt 'Patriots' and 'Jubilees' on express workings, 'Black Fives', 'Jubilees', unrebuilt 'Patriots' and Stanier and Hughes 2-6-0s on parcels and fitted freights. Local workings might provide the sight of 2-6-4Ts or 2-6-0s.

The LNWR, LMS and London Midland Region were concerned to maintain the status of the premier West Coast expresses such as the 'Royal Scot' and 'Midday Scot', heavy trains with an ample provision of restaurant cars. The 'Coronation Scot' was not reinstated after the war and it was the summer of 1957 before a limited load express ran once more between Euston and Glasgow Central. This was the 'Caledonian', made up to eight coaches and headed by a 'Duchess' Pacific. A second pair of 'Caledonian' expresses was added the following year but the early morning down and late afternoon up departures did not prove very popular. Eight coaches presented no problem for a Pacific but the 'Royal Scot' of the early 1950s loaded usually to 13 coaches of BR Standard stock. Other Western Division expresses were often very heavy and though the 'Merseyside Express' between Liverpool and Euston might be booked for 14 coaches, at busy periods its customary 'Princess Royal' Pacific might be taxed with one or two more.

The year 1959 marked the end of the Indian Summer for steam on the West Coast route. Some of the Stanier Pacifics had been repainted in maroon livery from 1957. This went well with coaches in the lined maroon livery which from 1956 had replaced crimson and cream (blood and custard to many enthusiasts!) except on the Western and Southern Region. The SR changed to unlined green paintwork to match its EMU stock. Although the WR had changed to chocolate and cream for its premier services, all other coaches were painted lined maroon.

The Eastern Region in the 1950s

The Great Northern main line began the decade with the introduction of standardised departure times for expresses leaving King's Cross: Scottish and Newcastle trains left on the hour, and Leeds trains at 18min past, although there were a few exceptions to this pattern such as the Pullman trains. At the start of the 1950s timekeeping of the expresses was far from good and the previous lengthy through workings were reduced to produce out and home trips for engines and men alike. There were extensive transfers of ER Pacifics between main sheds. The changed workings did not affect the summertime 'Capitals Express' (it was renamed the 'Elizabethan' in 1953 when the schedule was cut to 6¾hr) which continued as a through working using 'A4s' from King's Cross and Haymarket (Edinburgh) sheds. In the 1952 season just one engine failure was recorded on the 'Elizabethan', and Haymarket's No 60027 *Merlin* completed 42 return trips that year.

There was a move towards early morning departures for business trains, such as the 8am King's Cross-Leeds/Bradford introduced in the winter of 1952/3; the next September a Newcastle portion was added. This train was to become associated with the 'A1' Pacifics which normally worked it and put up some very fast runs. Although passenger motive power at the southern end of the GN main line did not change much through the 1950s, apart from short-term trials with an LMS 2-6-4T in 1954, and with BR and LMS 4-6-0s the next year, it was different on the freight side. In 1954, a batch of '9F' 2-10-0s replaced the Austerity 2-8-0s on the New England (Peterborough)-Ferme Park coal trains. Other '9Fs' followed later.

New from the winter 1956/7 timetable was a fast, limited load Anglo-Scottish express named the 'Talisman' whose journeys, like those of the prewar 'Coronation', left the respective terminals of King's Cross and Edinburgh Waverley in the late afternoon. South of Newcastle, the train was normally headed by an 'A4' provided either from King's Cross or Gateshead shed. A 'Morning Talisman' service was introduced in the summer 1957 timetable and, again, the usual motive power was an 'A4'. By now, double blastpipes and chimneys were being fitted to this class at a cost of £5,000 per engine (some 'A4s' had been built new with them) and some scintillating performances were recorded. The first main line diesels arrived at Hornsey shed in 1958, to signal the end of supremacy for the ER Pacifics although the unreliability of the diesels ensured that the steam locomotives were kept hard at work. Steam was also being replaced on GN suburban services, and from late 1958 new Type 2 diesels arrived to work outer suburban trains with diesel multiple-units taking over the inner suburban service. Yet another record achievement by an 'A4' came in May 1959 when a railway society special between King's Cross and Doncaster and back was hauled by No 60007 *Sir Nigel Gresley* which attained 112mph on the descent from Stoke Summit during the return run.

The 'Britannias' had their heyday in the 1950s when they were working Great Eastern line expresses. After the 1949 inauguration of the Liverpool Street-Shenfield electrification, the ER's train planners put their energies to redesigning the Liverpool Street-Norwich services via Ipswich and via Cambridge.

Below:
The 'A3' Pacifics were rejuvenated with the fitting of double blastpipes and exhausts. No 60062 *Minoru* brings an express for King's Cross past Woolmer Green signalbox, south of Knebworth, on 13 June 1959. *E. R. Wethersett*

Meanwhile, the Railway Executive was planning for the introduction of the BR Standard locomotives, including a mixed traffic Pacific. The Executive was acutely aware of the growing stock of Bulleid Pacifics and their likely underemployment on the Southern Region. There was a proposal to transfer 15 of these engines to the GE and, in April 1949, the first arrived.

Had the offer of the 15 Bulleid engines been taken up then the GE would have been written out of the allocation of the new Standard Pacifics. After trials with 'Battle of Britain' No 34059 it was returned to the Southern, and planning proceeded with the deployment of the Standard Pacifics to the GE. First reports were that 15 of the new Pacifics would go to the Eastern Region. The initial allocation was scheduled as Nos 70000-14. After its naming as *Britannia* on 30 January 1951, No 70000 was in service from Stratford shed shortly afterwards. Although promised to the ER, Nos 70004/14 passed to Southern Region stock.

The new timetable with 'Britannia'-worked trains between Liverpool Street and Norwich began on 2 July 1951, with all but hourly daytime departures from Liverpool Street, and from Norwich. The revised Norwich service was revolutionary in its overall approach to a steam-worked service for the timetable had been completely recast, new engines introduced, services uniformly accelerated and big savings achieved in the number of men, locomotives and coaches required.

Above:
Old and new in East Anglia! Waiting in the bay platform at Haughley on 23 July 1952 is the Laxfield branch train, headed by ex-GER 'J15' 0-6-0 No 65447. Speeding past with the down 'Norfolkman' express for Norwich is 'Britannia' Pacific No 70003 *John Bunyan*. The Laxfield branch was built by the Mid-Suffolk Light Railway and total closure came on 28 July 1952.
R. E. Vincent

Right:
The 1,500V dc overhead electrification dominates the scene at Stratford as an up Southend line train passes behind Gresley 'K2' 2-6-0 No 61721.
R. E. Vincent

The GE operators confirmed the Railway Executive's description of the 'Britannias' as mixed traffic engines for included in the Norwich and Stratford diagrams were newspaper, fish and goods trains. The BR Standard corridor stock was first placed in service during 1951 and a set of the new vehicles began running on the 'Norfolkman' from 7 May that year. A 'Britannia' at the head of the new coaches represented the truly BR standard train!

The Great Eastern had to wait until the summer 1953 timetable before there could be a similar reorganisation of the Cambridge line express service. Once again, the radical reshaping of the timetable brought appreciable savings in resources. It also benefited the Colchester main line as a result of the interworking of locomotives and stock. The 1953 changes were made possible with the arrival of a new tranche of 'Britannias'. Nos 70030/4-42 were immediately delivered to the ER, but Nos 70030/34 went first to the LMR, and then were loaned during May 1953 to Stewarts Lane to cover a temporary withdrawal of the Bulleid Pacifics. With their arrival on the ER there were now 13 Pacifics at Stratford, and 10 at Norwich.

The new Cambridge line timetable comprised a variety of trains, including Liverpool Street-Ely-Norwich expresses which were now handled by Pacifics. Several of the 'Britannias' diagrams called for an engine to complete two return trips between London and Norwich within 24hr or so. Not all the trains in the timetable were handled by the 'Britannias' and LNER-design 4-6-0s were used on the lighter jobs. Once again, regular interval departures applied from London, and from Cambridge in the up direction. Schedules were less dramatically improved than had been the case with the Norwich line.

In June 1956, came the announcement that the Colchester-Clacton/Walton line would be electrified at 25kV ac. There were also plans for the dieselisation of the GER, the first scheme for the delivery of 144 main line diesel locomotives being provisionally approved in late 1957. By autumn 1958, a much bigger scheme had been evolved to achieve the total elimination of steam traction in East Anglia. The ER announced that those 'Britannias' displaced from the Norwich service by the new diesels would be employed to speed up Liverpool Street-Clacton trains.

East Anglia had long been a home for aged locomotives and coaches working lightly used branch lines. Plenty of former Great Eastern locomotives, 2-4-2Ts, 'Claud Hamilton' 4-4-0s, 'J15' 0-6-0s and even the 'E4' 2-4-0s were to be seen, but steam-worked local trains were replaced from 1956 by diesel multiple-units. During 1958 diesel shunters arrived in East Anglia to put paid to ex-GE 0-6-0Ts.

By January 1959, most GE Line branch passenger services were dieselised and the service of through coaches from Liverpool Street to most coastal resorts now gave way to connections made by DMUs. Only the 'Broadsman' now worked beyond Norwich, to Sheringham and back. One interesting aspect of the use made of the 'Britannias' on the Liverpool Street-Norwich trains after 1958 was of their diagramming to

Below:
One of the weekday cross-country services introduced in early BR days was the Cleethorpes-Lincoln-Birmingham New St train which was worked throughout by an Eastern Region 'B1' 4-6-0. No 61157 was photographed leaving Derby Midland for Birmingham on 27 June 1959. *T. G. Hepburn/ Rail Archive Stephenson*

work alongside diesels. This lasted until the Pacifics left Norwich in 1961 and there was no other instance of similar scheduled steam/diesel working on BR.

The other big development in East Anglia was the closure of almost all the former Midland & Great Northern (M&GN) Joint system which in several places duplicated ex-Great Eastern through routes. It was almost entirely steam-worked. Most M&GN lines were loss-making and, in March 1958, the Eastern Region decided to proceed with a plan for closure of the major part of the system. This was implemented in February 1959 and apart from the line between Cromer and Melton Constable, all that survived were freight-only stubs of former through lines.

The ER operated another main line out of London, the former Great Central (GC) route through the East Midlands to Sheffield. By the 1950s the regular expresses were largely worked by 'A3' Pacifics or 'V2' 2-6-2s and the service as a whole was both smart and punctually operated. At summer weekends there were numerous holiday trains working over the GC from Sheffield and the Midlands cities, to and from the South Coast and Thanet resorts, and to Skegness, Mablethorpe, Blackpool and Llandudno.

The best-known freight service on the GC was that between Annesley and Woodford Halse, conveying coal to the south and west; by the mid-1950s '9F' 2-10-0s had taken over from ex-LNER and Austerity

Left:
East Midlands branches began to close in earnest during the mid-1950s. The service between Lincoln and Shirebrook North was one such example, and regular daily trains finished in September 1955. In April 1954, a Shirebrook North train leaves Lincoln on the former Lancashire, Derbyshire & East Coast line behind ex-GCR 'A5' 4-6-2T No 69812.
P. J. Lynch

Below:
The special train run on 20 September 1953 to commemorate the centenary of Doncaster Works was headed by ex-GNR Small Atlantic No 990 *Henry Oakley* (leading) and Large Atlantic No 251. The special is seen approaching Hatfield.
Ian Allan Library

2-8-0s. The GC was another of the lines to be affected by the redrawing of Regional boundaries, and from February 1958 the main line south of the Chesterfield area was transferred to the LMR. Before long, ex-LMS 4-6-0s were allocated to GC line depots and the Pacifics transferred away. A more dramatic development came from the decision to withdraw the through expresses between Marylebone and Manchester, Sheffield or Bradford. They ceased in January 1960, to be replaced by a limited number of semi-fast trains between Marylebone and Nottingham.

During the 1950s, the Great Northern section of the LNER gave a favourable reception to enthusiasts' requests to run special trains, and indeed put on some of its own to mark important anniversaries connected with the Great Northern Railway. In September 1953, the ER had agreed to the running of specials which celebrated the centenary of Doncaster Works, and more remarkably arranged for the return to steam of preserved Ivatt Atlantics Nos 990 *Henry Oakley* and 251. Apart from hauling the special trains the pair were turned out to work service trains from King's Cross to Peterborough on more than one occasion. In 1954, No 251 was back again on the main line, and

worked two special excursions, one to Liverpool and the other to Farnborough for the Air Show.

The North Eastern Region in the 1950s

What was distinctive about this Region? After all, it shared the working of the East Coast main line with the Eastern and Scottish Regions, had few express passenger trains of its own, and its coaching stock fleet was jointly owned with the Eastern Region. Based at the fine headquarters buildings of the former North Eastern Railway in York, the NER-controlled territory extended to Marshall Meadows, north of Berwick-upon-Tweed, south to Shaftholme Junction, just north of Doncaster, while its eastern boundary was the North Sea coast. As a result of changes in Regional boundaries in 1950 the Region included also the network of lines in Yorkshire's West Riding which once formed part of the GNR, L&YR, LNWR and

Below:
One of the British railway marvels of the 1950s was reckoned to be the York power signalbox, whose interior appears in this carefully posed picture. *Science & Society Picture Library 1690*

Above:
During the mid-1950s the North Eastern Region worked fast to
dieselise its local passenger services, or to withdraw them as
uneconomic. An up local train climbs away from Durham on
22 July 1958 behind ex-NER 'A8' 4-6-2T No 69883, one of a class
that had been rebuilt from 4-4-4Ts in LNER days.
D. M. C. Hepburne-Scott/Rail Archive Stephenson

Midland Railway. Away from the East Coast main line,
and the original limits of the North Eastern Railway,
the handover points with other Regions were at
Bolton upon Dearne, Cudworth, almost to Barnsley
and Penistone - all with the Eastern Region, and just
to the east of Skipton, at Hebden Bridge, and to the
east end of Standedge Tunnel, with the LMR.

The Region's base was the industrial strength of the
north-east. The freight workings in this area were
largely in the hands of ex-North Eastern Railway 0-6-0s
and 0-8-0s which had changed little from the 1920s
and went on beyond the demise of the North Eastern
Region itself. One of the Region's showpieces involved
the annual movement of some $1\frac{1}{4}$ million tons of
imported iron ore from Tyne Dock to the steelworks at
Consett in north-west Durham. Block trains of special
56-ton hopper wagons were used, and it was awe-
inspiring to travel up the steeply graded line to Consett
with '9F' 2-10-0s, one at the front of the train, and one
banking at the rear. The '9Fs' had replaced 'O1' 2-8-0s
and ex-NER 'Q7' 0-8-0s. There was also the enormous
traffic generated by the chemical industries on
Teesside, so it was not surprising that nearly 80% of
the NER's revenue came from freight.

The NER was proud of its section of the East Coast
route, and particularly with the Darlington-York high-
speed section which was largely four-tracked. It was
also easily graded and saw some of the fastest steam-
worked express trains of the 1950s. In 1952, the
Region boasted no less than three of the speediest

point-to-point runs on the whole of BR. Two of these
were made by the 'North Briton' Leeds-Glasgow
express, the southbound train covering the 44 miles
from Darlington to York at an average speed of
62.9mph, the fastest run on BR.

The NER built on pioneering work in signalling
technology by the North Eastern Area of the LNER.
The route relay interlocking installed at the Thirsk,
Northallerton and Darlington signalboxes had been
notable in late prewar years, but 1951 saw the
commissioning of the all-electric signalbox at York
which for a time was the largest of its type in the
world. It controlled over 33 route-miles, and with its
completion, colour-light signalling extended for more
than 50 route-miles along the East Coast main line.
Another modern signalbox was installed later in the
1950s at Newcastle which was the Region's busiest
passenger station.

Away from the trunk routes, the Region served
miles of attractive countryside in the north-east. There
was the cross-country main line, from Newcastle to
Carlisle. There was also the extraordinary and now
largely dismantled line from Darlington and Barnard

Castle across the high Pennines by way of Stainmore Summit (1,370ft above sea level) to Kirkby Stephen, where the line diverged to serve Tebay in one direction, and in the other joined the West Coast route south of Penrith. The main section of line included the 200ft high and 1,000ft long Belah Viaduct. Other than at weekends, the line's passenger traffic was meagre but its *raison d'être* was coke traffic passing along it from the north-east to supply the Furness steelworks. The line was expensive to work. Heavy locomotives were prohibited and so three 2-6-0s or 0-6-0s were required to work coke trains loaded to the maximum 34 wagons up the 1 in 60-odd gradients to Stainmore. Other remarkable and difficult to operate lines were along the Yorkshire Coast from Bridlington to Scarborough and on to Whitby and Middlesbrough, with viaducts, steep gradients as severe as 1 in 39 and frequent curvature.

Whether here or on its busy passenger routes in urban areas, from the early/mid-1950s the NER forged ahead to obtain authorisation for schemes to replace steam trains by diesel units. By 1956, regular steam passenger trains had largely disappeared from the Newcastle-Middlesbrough and Newcastle-Carlisle services, and by 1959 DMU operation had eliminated steam from local passenger workings on many of its other routes.

In 1956, about half of the NER's 1,500-strong steam fleet was composed of former North Eastern Railway engines, the most numerous being 120 'Q6' 0-8-0s, 115 'J27' 0-6-0s, over 80 'J72' 0-6-0Ts (some of which had been built as late as 1951) and 69 'B16' 4-6-0s. Few ex-NER passenger tender engines remained - just 14 'D20' 4-4-0s.

Gateshead, Heaton, York and Leeds Neville Hill sheds together had nearly all the NER's allocation of 70 LNER-design Pacifics, and there were 65 'V2s' on the Region, over 80 each of the LNER 2-6-0s and 'B1' 4-6-0s and 50 or so Gresley 'D49' 4-4-0s. The locomotive allocation was rounded off by numerous ex-LMS and BR Standard classes and Austerity 2-8-0s. York shed was celebrated for its large allocation of engines - the total was 164 in 1956. One unusual feature of the NER was the opening in mid-1958 of a new steam depot at Thornaby which had an allocation of 148 engines and replaced two older depots on Teesside.

The Scottish Region in the 1950s

English railway enthusiasts were always eager to see Scottish Region (ScR) locomotives south of the Border and their appearance in the London area was definitely unusual, if not rare, although in summer there was the regular through working of a Haymarket 'A4' on the 'Elizabethan' express. Tell-tale signs of Scottish motive power were the light blue painted smokebox numberplates, a feature of some Scottish sheds such as Polmadie and, in the case of engines shopped at St Rollox Works (Glasgow), larger than normal cabside numerals.

The ScR purchased three steam locomotives in 1957, three ex-War Department Stanier '8F' 2-8-0s at a cost of £16,720 including conversion from oil to coal burning. The engines became BR Nos 48773-75 and their purchase was justified on account of additional iron ore traffic to Colville's. During 1957/8, the ScR was still receiving new BR standard engines, including Caprotti-fitted '5' 4-6-0s and also Class '4' 2-6-0s.

Left:
Many of the Scottish Region sheds made a real effort to keep their engines looking smart. Look at this well-groomed LMS 'Black Five' 4-6-0, No 44994, working a Glasgow Central-Edinburgh Princes Street stopping train near Uddingston on the exit from Glasgow, 24 July 1952. Into the 1950s the semaphore route indicators employed by the former Caledonian Railway continued to appear on engines, as displayed above the buffer beam of No 44994. *E. R. Wethersett*

Above:
Nearing the end of regular steam operation on the Far North line from Inverness-Wick/Thurso. At the head of the 11.5am from Inverness on 8 September 1959, LMS '5' 4-6-0 No 45496 takes water at Helmsdale before depositing the former Highland Railway TPO van (foreground) and the train's restaurant car in a siding to await a southbound working. Then No 45496 and remaining vehicles will set out for Wick. *Brian Stephenson/ Rail Archive Stephenson*

The general impression was that the Scottish Region kept its engines in better external condition than the other Regions. Once north of the Border the BR vitrolite totem signs proclaiming station names (and now changing hands at auctions for exorbitant prices) were in a distinctive and cheery light blue with white lettering. Stations were painted a variety of colours, although a cream and brown scheme seemed to be common. The largest Scottish stations had a dignity all of their own: Edinburgh Waverley and its acres of glass roofing was in a class of its own and the sheer scale of Glasgow Central was awe-inspiring. Glasgow St Enoch's graceful roof and delightful tea-room were refreshing, Perth held the promise of the Highlands and the Far North, while Aberdeen was truly the Gateway to the North. Having

reached either Perth and Aberdeen, it was only then that the traveller appreciated that still more railway beckoned him on - the Far North line extended 161 miles to Wick from Inverness.

Then there were the many branches - a real delight, whether the Ballachulish line from Connel Ferry and its Caley 0-4-4Ts; the Killin Junction-Killin branch with its single-coach trains (some were mixed trains); the Tillynaught-Banff line, or that from Craigellachie-Boat of Garten. Never mind the passenger traffic on the last-named, it served a number of whisky distilleries, some of which had their own beautifully turned-out steam shunter.

The Craigellachie-Boat of Garten line along Speyside was worked by ex-Great North of Scotland 'D40' 4-4-0s until late in the 1950s, the last being No 62277 *Gordon Highlander* which survived in service into 1958. It was then restored to GNSR green livery (which in fact it had never carried) and worked specials on the main line. This was to the credit of the ScR although, to be fair, its officers had changed their minds over the preservation of the 'D40'. In late 1957, the Scottish Area Board had decided that 'no action should be taken to preserve *Gordon Highlander*'; pleas from the public saved the engine.

A rather different initiative came in March 1958 when preserved Caledonian Railway Single No 123, together with two restored CR coaches were used for a series of public excursions. This was the prelude for a more ambitious exercise in September 1959 when the Scottish Industries' Exhibition was held in Glasgow and the ScR decided to run a number of public excursions from various parts of Scotland in connection with the show. The tally of attractive restored engines included No 123, the restored ex-GNSR 4-4-0 No 49 *Gordon Highlander*, ex-North British 'Glen' 4-4-0 No 256 *Glen Douglas* and Highland Railway Jones Goods 4-6-0 No 103. Also in 1959, two ex-North British 'D34' 4-4-0s made a fine sight at the head of special trains on the West Highland line.

Talking of which, the LNER beaver-tail observation cars formerly used on the 'Coronation' finally found regular work on the West Highland line from 1956; the cars were joined by an ex-'Devon Belle' car during 1958 and then the next year the two LNER vehicles were partially rebuilt to provide better vision for passengers. All three cars were employed on Glasgow-Fort William, Fort William-Mallaig and Glasgow-Oban trains in summer. The 'Devon Belle' car was later used on the Inverness-Kyle of Lochalsh line.

The early 1950s saw sweeping withdrawals of passenger services from ScR branches. In 1951 alone, almost 50 branch lines totalling 430 route-miles lost their services. Many of these lines were to survive 10 years or more with freight services only. However rural and remote, some of these lines carried heavy seasonal traffic such as seed potatoes, in addition to the regular business in general agricultural products, grain inwards to the distilleries and whisky outwards, and domestic and industrial coal. Fish traffic kept the Peterhead and Fraserburgh lines busy.

In 1958, it was agreed to remodel the passenger services between Inverness and Wick/Thurso which involved the closure of a number of intermediate stations, as well as The Mound-Dornoch branch. This little line had achieved fame for employing the last Highland Railway engines in BR service - the tiny 0-4-4Ts Nos 55051/53, the latter being withdrawn in January 1957. The replacement was totally unexpected - a WR '16xx' 0-6-0PT, Nos 1646. It was joined the following year by No 1649 and both continued at work in the Far North until the Dornoch branch was closed completely in June 1960.

Construction of the Alford branch, extending 16 miles from Kintore on the Aberdeen-Keith line into Donside, had been independently promoted, but by 1866 had been amalgamated into the GNSR. The branch did much to develop local granite quarrying. The line was popular for excursion traffic but its passenger service was never extensive. Steam rail-motors were used briefly. In 1910, there were just four trains each way.

In the November 1949 *Bradshaw* the Alford branch passenger trains amounted to three trains each way and the average journey to and from Aberdeen took 1¼hr for the 29 miles. Lines such as this could not compete with buses and the Alford branch closed to passenger traffic in January 1950, but it lingered on until 1966 in freight use. Today the station building at Alford is a transport museum and a miniature railway runs from there to Bridge of Alford.

One of the branch lines in the Border Country was the 10½-mile line opened in 1901 from Fountainhall on the Waverley Route to the town of Lauder. It had been constructed under the Light Railways Act of 1896 but, apart from trout fishers, was seldom patronised by passengers, the service being withdrawn as early as September 1932. Probably the line would have attracted scant attention from people interested in railways had it not been for the motive power used after 1944. It was worked with an ex-Great Eastern Railway 'J67' 0-6-0T which trailed an ex-North British tender containing the water supply. The engine's side tanks remained empty, for a 'J67' with full tanks would have been too heavy for the lightly-laid line. Its complete closure came in 1958 and an Ivatt 2-6-0 had earlier displaced the GE tanks.

The Waverley Route closed along nearly all its length in 1969 although there are currently plans to reopen a railway through the Borders. This romantic line was once an Anglo-Scottish trunk route, used for the Midland and North British companies' jointly-worked service from London, the East Midlands, West Riding and Carlisle to Edinburgh and points north thereof. After 1923, the LNER and LMS co-operated to provide the service which used joint coaching stock into the 1930s when the line's role as a trunk route began to diminish.

The 98-mile railway passed by way of Hawick, St Boswells and Galashiels. A journey over the line was an unforgettable experience, given the progress through countryside rugged, bleak and beautiful by turns, combined with heavy gradients, curvature, and speed restrictions. Much of the line's activity in later years came from through freight traffic passing between the marshalling yards at Carlisle and Edinburgh. The effect of heavily laden van trains rushing through stations such as St Boswells in the still of an evening was quite dramatic.

The weekday passenger service of the 1950s comprised the through daytime 'Waverley' express each way between St Pancras and Edinburgh Waverley via Nottingham, Leeds, the Settle & Carlisle line and Carlisle; listening to the station announcer at St Pancras intoning the list of Waverley Route stations the express served was an experience in itself. There was also an overnight sleeping car service each way between St Pancras and Edinburgh. The remainder of the through trains between Edinburgh and Carlisle were semi-fasts, calling at a score of stations and taking up to 3¾hr on their journey.

During the 1950s the through trains were worked by the various classes of LNER Pacific and 'V2' 2-6-2s.

Those trains going no further south than Hawick were likely to be powered by 'B1' 4-6-0s. All three types of engine plus 'K3' 2-6-0s and LMS '5' 4-6-0s were used on through goods workings.

At St Boswells, connection was made by the branch to Kelso and on to Tweedmouth (the passenger trains continued to Berwick-upon-Tweed). Meandering through some beautiful scenery, this line remained steam-worked as late as 1963, almost as if its existence had been forgotten by the railway planners. In the 1950s, either an ex-North British 4-4-2T or an NB 4-4-0 was the customary motive power, but Class '2' 2-6-0s later took over.

This sketch of the Scottish Region of the 1950s is of course selective. For a start, it ignores the busy inter-city express services between Edinburgh and Glasgow, Edinburgh and Aberdeen, Glasgow and Aberdeen, Glasgow/Edinburgh and Inverness, not to mention the Anglo-Scottish sleeping car trains. It has necessarily turned a blind eye to the heavy suburban traffic of Edinburgh and Glasgow, and the intensive freight working in the Lowlands and the Kingdom of Fife.

Below:
Well, not every Scottish engine was spick and span! Rather woebegone 'A3' Pacific No 60096 *Papyrus* crosses Tay Bridge with the 11.50am Dundee Tay Bridge station-Edinburgh Waverley on 23 July 1952. *E. R. Wethersett*

6. The 1960s:
The Steam Railway Bows Out - Oliver Cromwell Sounds the Final Whistle

O n New Year's Day, 1960, few could have imagined that BR standard gauge steam would not survive the decade. But one Region was working to a plan - even if it was never made public. The Western Region (WR) had agreed with the British Transport Commission as early as October 1956 that the complete conversion of the Region to diesel traction, area by area, would be achieved by the end of 1968 using 1,425 main line diesels in the process. The two years 1955 and 1956 were indeed crucial in the decision to press ahead with displacing steam. The British Transport Commission had published its Modernisation Plan in 1955, and the first Pilot Scheme diesels were ordered later that year. These were the prototypes with which BR intended to gain experience before standardising the designs for mass production. It was a sound principle but one that was ignored: before any such data had become available, large orders were placed for largely untried classes of main line diesels.

Following publication of the Modernisation Plan, the proposed 1956 Building Programme for steam locomotives was slashed. The total of 263 engines struck out from the 1956 Programme included 36 'Britannia' Pacifics for three Regions, 20 'Clans' for the

North Eastern Region, and the first of a new Class '8' 2-8-0 design. With curtailment of the 1956 Building Programme the remaining BR Standards to be built were 84 '9Fs' and 61 mixed traffic engines authorised as an additional build to the curtailed 1956 Programme. Swindon Works took longer than expected to complete its order which is why the last of the batch, No 92220 *Evening Star*, came to be completed as late as March 1960.

The serious downturn in BR's revenue in 1958 partly reflected a recession in heavy industry but freight traffic was also being lost to road. BR's deficit for the year was £85 million. BR was committed to break even by 1962 and, with every indication that this was an impossible target unless urgent action was taken, the BTC pressed BR hard to make all-round economies including train service cuts, closures

Above:
The 1960s brought the end of the steam age railway in Britain. It is well summed up in this picture of Laisterdyke Sidings, near Bradford, on 2 September 1967 when steam was near to being eliminated in West Yorkshire. Alongside the yard is a passing summer Saturdays express from Bradford Exchange to Skegness, worked by the now preserved 'B1' 4-6-0 No 61306. In the yard is a loaded coal train ready to depart behind a Stanier '8F' 2-8-0.
Malcolm Dunnett

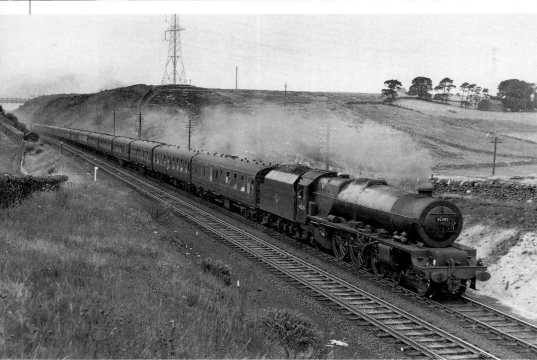

and reductions in staffing. Modernisation and rationalisation of the railways were seen as the key to BR's chances of improving its finances. Electrification and dieselisation programmes were speeded up, in particular the electrification of the LMR main line from Euston to the Northwest. Diesel locomotive production was stepped up.

The BTC carried out a review of the 1955 Modernisation Plan late in 1959 and forecast that passenger and freight rolling stock fleets would shrink in line with BR's reshaping as a 'considerably more compact railway system', to use the official description. One casualty of the review was the electrification of the GN main line to Leeds which was dropped without comment during 1959. Even so, despite a wide-ranging investigation of BR's operations there was no definite forecast for the end of steam. All that was said was that by 1963 more than 75% of BR's passenger train mileage would be worked by diesel or electric traction.

Nothing that happened in the first two years of the 1960s gave any impression of changes in overall policy. It was a time of the completion of a number of large dieselisation and electrification projects - the introduction of diesel multiple-units on the principal trans-Pennine routes, and electrification at 25kV ac of Great Eastern lines in northeast London, of the London, Tilbury & Southend section, of the Crewe-Manchester/Liverpool routes and of much of the Glasgow suburban services, not to mention completion of the SR's main line third-rail electrification in Kent. Diesel multiple-units replaced steam on other suburban services in the London area, between Manchester and Liverpool and elsewhere. The years 1960-62 saw main line diesels rapidly take over a number of the most important services and in time for the summer 1962 timetable when accelerated services were being offered which for the first time brought better running times on some routes than had applied in 1939. On the East Coast route 'Deltics' were now at work on the principal Anglo-Scottish and West Riding expresses.

To the general observer BR still appeared to be largely steam-worked, even if by 1962 relatively few weekday express passenger trains remained steam-hauled. Figures published for the first half of 1962 showed that steam haulage accounted for nearly 50% of BR's train-mileage. In June 1962 10,956 steam locomotives were on the books compared with 3,436 diesels - 1,956 of these being shunting locomotives - and just 169 electric locomotives. Nor did these figures reveal just how much of BR's passenger services had switched to DMU and EMU operation.

Virtually all improvements to steam locomotives ceased during 1962 so that the equipment of WR 'Castles' and LNER Pacifics with double blastpipes and chimneys came to an end. The rebuilding of Bulleid Pacifics had already ceased. One minor development was the fitting of 'Battle of Britain'

No 34064 *Fighter Command* with a Giesl ejector in May 1962; this was justified on the grounds of reducing spark emission.

One of the ways by which steam working was drastically reduced in the next few years was the consequence of closures and cut-backs to services. During 1962, the Western Region drastically thinned-out the Bristol area's largely DMU-worked local and suburban passenger services, so that units were released to displace steam-hauled trains elsewhere. There were piecemeal closures of lines, the heaviest programmes being pursued by the Western and Scottish Regions. The WR had undertaken a thorough study of its operations during 1962 and the upshot was a lengthy list of line closures, particularly in the West Country and Mid and West Wales; many of these were effected by the end of that year. Some of the doomed lines were little known outside their immediate locality, such as Wrexham-Ellesmere which lost its passenger trains on 8 September 1962. On the snowy 31 December of that year, the lines from Pontypool Road to Neath, Brecon-Moat Lane, Brecon-Hereford and Brecon-Newport, and a number of the area's branch lines all shut up shop. Over the previous few months enthusiasts had travelled extensively to try to obtain a last ride over these and many other routes.

The LMR was working to its Passenger Plan which had been drawn up early in 1959; this report

Below:
A farewell to steam at Crewe Works. The ceremony at the works on 2 February 1967 when 'Britannia' Pacific No 70013 *Oliver Cromwell* was about to be returned to traffic. It was the last general service steam locomotive to receive a heavy repair at BR workshops. *John R. Hillier*

commented that 'recent signs of changes in Ministerial policy make it possible for passenger services to be looked at more objectively'.

Despite the momentum of the Modernisation Plan with its heavy investment in new equipment, it was impossible to disguise BR's losing battle against its competitors. Traffic fell badly in 1961 and BR's deficit increased. Government's scrutiny inevitably homed in on the management of nationalised transport. The old British Transport Commission was at last - and far too late in the day - regarded as superfluous. It was criticised for being far too concerned with BR's problems, to the disadvantage of its other charges - London Transport, British Transport Docks, Inland Waterways and the Transport Holding Company (concerned with bus and road freight operations). Freed from BTC control, these concerns now became separate entities and the railways were now run by the British Railways Board (BRB) which began operations on New Year's Day, 1963. Its Chairman was one Dr Richard Beeching. He was an industrialist who had been called in by Government over the last four or so years to look at the railways' problems and from 1961 he had been Chairman of the BTC.

By March 1963, Dr Beeching had published the famous report associated with his name but whose title was actually *The Reshaping of British Railways*. The lists of lines and stations for closure, together with the packet of maps accompanying them, are what first come to mind. In fact, the report consisted mostly of an analysis of the traffic actually handled by the railways, whether or not it was profitable, and how the best-paying flows might be developed to become more lucrative. The report summarised its own content as the investigations carried out, the conclusions which were drawn, and the proposals which were made 'for the purpose of reshaping British Railways to suit modern conditions'. 'The railways,' it continued, 'should be used for purposes for which they offered the best available means … and 'they should cease to do things for which they are ill-suited.' The answer, said *The Reshaping of British Railways*, was to concentrate on traffic 'carried in dense flows of well-loaded through trains'.

The Report was better on analysis than prescription but then the BRB expected the railway Regions to carry out its recommendations for the Board was not concerned with hands-on management. Some of the Regions had anticipated some of what the Report said, and were already cutting back dead wood and concentrating on developing the best-paying traffic. The process needed to be even more energetically pursued but it would be several years before the BRB exercised stronger control. When it came to motive power, admittedly not a major concern of

The Reshaping of British Railways, it was surprising none the less to find that of 15 recommendations to achieve the reshaping of the BR, the continued replacement of steam by diesel locomotives was rated 14th in the list.

The LMR's withdrawal of BR's last standard gauge steam locomotives in the summer of 1968 was achieved slightly earlier than expected. The very last movement was that of No 70013 *Oliver Cromwell* which had been one of the engines selected to work BR's last steam special on 11 August. It worked light engine overnight from Carlisle to Wymondham in Norfolk, in preparation for being placed in the museum at Bressingham. When *Oliver Cromwell* sounded its whistle for the last time in BR service, on the morning of 12 August 1968, this was the full stop in a story that had started over 140 years before.

The Western Region in the 1960s

At the start of the 1960s, many parts of the Region were still essentially GWR. Trains of Great Western stock set off into rural England or Wales. For instance, a 2-6-0 at the head of a Taunton-Barnstaple train would pass stations with lovingly tended gardens where, as likely as not, station staff would open carriage doors for boarding passengers. In GWR coaches dating from the 1930s, the photographs above the seat-backs were of Torquay in the days before its roads were clogged with traffic, or of Cheltenham Spa in the palmy 1920s. The map in between the pictures was defiantly headed Great Western Railway.

Of course it could not last. The Western Region's annual deficit was increasing alarmingly. Although many lines had been proposed for closure before the publication of the Beeching Report, the process accelerated once it had appeared. A significant proportion of the Region's network was closed to passenger traffic by the mid-1960s, much indeed closed entirely. Lengthy cross-country lines such as Carmarthen-Aberystwyth (56 miles long) were cleared from the maps. Public opposition to closures ensured however that a railway has survived to cross central Wales into the 1990s, this being the former Craven Arms-Llandeilo line of the LNWR although the facilities at stations and signalling were rationalised.

Opposite top:
Small chance of custom at Llanarthney, on the former LNWR Llandeilo-Carmarthen line. A one-coach train is worked by a WR '74xx' 0-6-0PT on 3 September 1963. *Andrew Muckley*

Opposite:
'54xx' 0-6-0PT No 5416 leaves Yeovil Pen Mill station with an auto-train to Yeovil Town on 23 June 1963. *G. D. King*

DMUs replaced regular steam-hauled passenger trains in the summer of 1964, and all through freight workings ceased.

Brecon was the best known perhaps of those county towns served by the WR to have lost their trains; it was said that on a typical day in the early 1960s only 200 people joined trains serving the town station. Another county town to lose its railway in 1962 was Cardigan. Travel over the lightly engineered 27-mile railway from the main line junction at Whitland was to be relished. The timetable sternly informed - Weekdays Only - Second Class Only. One of the GWR's '45xx' 2-6-2T would scuttle round the curves during the 90min journey, with the train calling en route at delightfully named halts such as Login, Glogue and Boncath. On the 6.50am from Cardigan you might find no more than a single coach headed by one of the small '16xx' pannier tanks.

Some of the WR lines retained their through services, such as that through the Golden Valley down from Sapperton to Stroud and beyond - the Swindon-Gloucester line. This had lost the Gloucester-Chalford local service from 2 November 1964, worked to the end by auto-trains consisting of a '14xx' 0-4-2T, and at least one GWR-pattern auto-trailer. Originally worked by a steam rail-motor, trains called at a series of halts - among them Ebley Crossing, Cashes Green, Downfield Crossing, Bowbridge Crossing, Ham Mill - some clue to the valley's mills and its hamlets.

If 1962 saw the loss of a number of branch lines, a tidying up process involving altered boundaries between BR Regions was implemented on New Year's

Above:
Expensive operations on the Central Wales route. A Stanier '8F' 2-8-0 at front and rear drag a northbound freight towards Sugar Loaf Summit on 30 October 1961.
B. J. Ashworth

Right:
Economies were sought on the Cambrian lines and a severely rationalised network emerged from the 1960s. '2251' 0-6-0 No 3200 passes Talerddig with a down freight train on 11 May 1961. *J. R. Besley*

Day, 1963. This resulted in the LMR gaining all ex-GWR lines north of Aynho Junction (which is just south of Banbury) including the Birmingham and Wolverhampton areas, and all North Wales served by the Cambrian and Ruabon-Dolgellau lines. The WR's largest centre in the north of its much reduced territory was Worcester. Meanwhile, from the same date the WR gained former Southern Region lines west of Salisbury, principally the West of England main line to Exeter and its branches, and the 'Withered Arm' into North Cornwall and the Exeter-Barnstaple-Ilfracombe line. The object of this change of control was to rationalise the West Country network.

The traditional pattern of services was changing rapidly. In September 1962, the 'Western' class diesel-hydraulics had taken over from 'Kings' on the Paddington-Birmingham-Wolverhampton expresses. Now it was being made clear that once the Euston-Birmingham-Wolverhampton service was electrified, the former GWR Birmingham main line would see semi-fast trains only, which would run no further north than Birmingham. The former through coaches from Paddington to the Cambrian Coast, to Shrewsbury, Chester and Birkenhead would be withdrawn in favour of DMUs connecting at Wolverhampton High Level station out of LMR electric expresses.

With the LMR's control of GWR lines in the Black Country, closures soon followed. With the withdrawal of through trains north of Birmingham Snow Hill,

before long - but after the end of steam in the area - the Birmingham-Wolverhampton main line was closed to passenger traffic. On the northern portion of the Oxford, Worcester and Wolverhampton (nicknamed the Old Worse & Worse) main line, the stopping service from Stourbridge Junction to Dudley and Wolverhampton had been discontinued earlier than that, having remained largely steam-worked to the end which had come in July 1962.

Some WR through routes had gone by default, most notably that between Lydney Junction, Sharpness and Berkeley Road which was on the former Midland Railway trunk route between Birmingham and Bristol. The Lydney Junction-Berkeley Road line included the famed Severn Railway bridge which had been built in 1875 by an independent company, with the support of the Severn & Wye Joint Railway and the Midland Railway. The bridge was nearly a mile long, and the single line of railway was carried 70ft above the high-water level of the Severn. The Company that caused the bridge to be built was later purchased jointly by the GWR and Midland. The bridge was built earlier

Below:
Confusing? This is Horfield, on the exit from Bristol by way of the former Midland Railway main line. At the head of Southern Region green-liveried coaches is 'B1' 4-6-0 No 61393 which is working a holiday express to the Midlands on 17 August 1963. Eastern Region 'B1s' from Sheffield area sheds regularly worked to the Bristol area on such trains in the early/mid-1960s. *W. L. Underhay*

Above:
Former Midland line, but ex-GWR motive power. From 1958, the WR had taken control of the Birmingham-Bristol line south of Barnt Green and the loco depots along this section. At Bredon, north of Cheltenham, on 29 August 1964 a freight travels northwards, perhaps for Worcester, behind 'Grange' 4-6-0 No 6815 *Frilford Grange. Derek Cross*

than the Severn Tunnel and, when the latter was under repair, trains were diverted by the Lydney Junction-Berkeley Road line, then via Bristol rather than taking the old GWR South Wales route via Swindon, Gloucester and Lydney.

The Severn Railway bridge was subjected to special trials in 1956 to see if 'Castles' could work across it safely, a 2-6-0 being the largest engine otherwise permitted. But the life of the bridge came to an abrupt end in October 1960 when it was struck by tank barges on the Severn, and in due course the remains of the structure were dismantled although some parts lingered until 1970.

With the changes affecting the WR's extent, steam was withdrawn sooner than planned. The fleet of express locomotives was decimated in 1962/63. Some of this was to be expected, with the dieselisation of the Birmingham main line expresses, but at the end of the summer 1962 timetable no less than 169 engines were condemned. They included 25 'Castles' and nine 'Kings' but somewhat inexplicably this cull included 'Castles' that within the previous year had received heavy overhauls when they had been fitted with double chimneys. Some of those 'Castles' left in service were in poor condition and deteriorated even further in their remaining two or three years' life. Once proudly named *Winchester Castle*, No 5042 was a wretched sight, filthy dirty, with nameplates and numberplates removed, bent handrails, battered chimney and missing safety valve bonnet.

GWR classes disappeared very quickly from view. The first 'Manor' was withdrawn in April 1963 and the class was extinct by the end of 1965. The rundown of Western steam and its replacement by diesel-hydraulic, and increasingly diesel-electric classes, continued through 1964 and the following year. By the autumn of 1965 most of the remaining ex-GWR engines on the WR were allocated to sheds in the London Division, including Southall and Oxford. Another base was Gloucester.

As late as December 1965, no less than 50 steam locomotives were observed in steam at Oxford shed, some being visitors from other Regions. One of the last expresses diagrammed for an ex-GWR engine was the Poole-York through train, normally headed by a 'Hall' between Oxford and Banbury. Despite being officially withdrawn, No 6998 *Burton Agnes Hall* was provided for the last day of WR main line steam on 3 January 1966; having achieved this distinction, it was only fitting that it should be purchased for preservation by the Great Western Society.

Not all GWR engines were taken out of service with the end of WR steam in January 1966. Apart from a pair of pannier tanks allocated to Bath for work on the Somerset & Dorset, some of those engines that had passed to LMR stock with the redrawing of Regional boundaries lasted into 1966. These included a couple of '16xx' 0-6-0PTs, and several '57xx' 0-6-0PTs, the last four of the latter being withdrawn in November 1966.

During the mid-1960s, cross-country expresses seemed to be regarded by BR as less than desirable, and little was done to make them attractive to users. They retained the interest of enthusiasts because the Poole-York train (routed via the Great Central) and the 'Pines Express' (Poole-Manchester/Liverpool) were steam-worked for parts of their journeys. Bulleid Pacifics, usually 'West Countries' or 'Battle of Britains', but sometimes a 'Merchant Navy', were used from Poole to Oxford and back into 1966. For a short while the SR Pacific on the York train worked through to Banbury, this shed providing an LMS 'Black Five' for its share in the working.

The 'Pines Express' had been rerouted with the closure of the S&D line to long-distance trains in 1962 and now ran from Oxford to Crewe via Birmingham and Market Drayton. When the GC main line was shut down during 1966, for a while the Poole-York train was rerouted via Worcester to Birmingham New Street, and then on to the Midland route to Derby and beyond.

Although the WR had planned to be diesel-only from the New Year of 1966, its plans were thwarted because the intended closure of the Somerset & Dorset line to passenger traffic could not go ahead until licences had been issued for the replacement bus services. Nine engines were allocated to Bath Green Park shed by the WR (SR engines worked through from the Bournemouth end) - comprising four Stanier '8F' 2-8-0s, two LMS '3F' 0-6-0Ts, two GWR '57xx' 0-6-0PT, and a BR '5' 4-6-0.

The demise of the S&D was set in motion once the WR had taken over most of the line in 1958 and had soon diverted through freight traffic via Bristol. Then the popular weekend excursions were mostly taken off. In 1962, the year-round 'Pines Express' and summer weekend through holiday trains were rerouted away from the S&D. In place of the indigenous engine classes, some WR motive power was brought in, and of course the BR '9F' 2-10-0s which performed so magnificently on the 'Pines' and other expresses. During 1964/65 the cut-backs included closure of the line at night (to save the cost of night-shift signalmen) and the closure of most local stations to goods traffic; some stations were downgraded to unstaffed halts.

During mid-1965 BR announced that the S&D was to be closed to passenger traffic but inevitable objections from users protracted the process. Late 1965 and the Minister of Transport agreed to the closure plan, announcing that the line would lose all its trains from early January 1966. The problem with licensing the alternative buses led to BR being forced to operate for two months longer. Costs were cut by running a skeleton train service which came to an end on the weekend of 5/6 March.

Below:
Nearing the end of the Somerset & Dorset line: on 11 December 1965, BR '4' 4-6-0 No 75072 waits at Radstock with the 16.35 Bath Green Park-Templecombe. *D. H. Aldred*

The Southern Region in the 1960s

By 1960, much of the Region's forward planning was on course. With the completion of Phase 2 of the Kent Coast electrification, combined with the acquisition of a fleet of straight electric, electro-diesel and diesel-electric locomotives, steam would be eliminated from the South Eastern Division (as the former Eastern Section was now called). The final date for electrification had been set as 1962 but June 1961 saw the end of main line steam working on the South Eastern Division, symbolised perhaps by the arrival on Sunday, 11 June at London's Victoria station of the final steam-hauled up 'Golden Arrow' behind rebuilt 'Bulleid Pacific' No 34100 *Appledore*.

What about the rest of the SR network? Under the 1955 Modernisation Plan, the then Eastern and Central Divisions were to be electrified and steam working eliminated. Electrification would have been extended to the Allhallows and Grain branches and to the Oxted lines, including as far south as Horsted Keynes, and also have included the lines from Guildford to Reigate, Wokingham Jct-Aldershot South Jct lines and Christ's Hospital-Shoreham. In advance of electrification they would be dieselised. By 1962, however, it was decided that the electrification of none of these lines could be justified.

Above:
Steam haulage of the 'Golden Arrow' was soon to cease when rebuilt 'West Country' Pacific No 34100 *Appledore* eased the down train towards Dover Marine on 30 May 1961.
Rev A. C. Cawston

A survey in 1955 had shown that a number of the Central Division's lines were unremunerative but it was argued that their dieselisation would enable a better service to be provided more cheaply. By 1959 it was decided that the increasingly busy Oxted lines would be dieselised before electrification took place in the mid-1960s, using units that had been intended for later dieselisation projects. The continued steam working of the Oxted lines had led to increasing public complaints, said the SR, and the arrival of a limited number of diesel sets would allow steam traction to be confined to a few trains only. Later, diesel locomotives were used to work the peak-hour business trains but it was 1964/65 before the main Oxted services were free of steam, and the following year before the remainder of the associated branch lines, such as Eridge-Polegate, were either dieselised or lost their passenger trains altogether.

The promise in the Modernisation Plan to use main line diesel locomotives on the Waterloo-Bournemouth/Weymouth and Waterloo-Salisbury-

Exeter routes seems to have been dropped soon after 1955 because the next approach was to look at electrification of the Waterloo-Bournemouth line. In the meantime, Bulleid Pacifics, BR Standard and members of some Southern Railway locomotive classes made redundant by the electrification and dieselisation of the South Eastern and Central Divisions were transferred to the South Western Division sheds to displace many of the remaining pre-Grouping engines. Also, as many as 28 Type 3 diesel locomotives were spare from the Kent Coast scheme and these were redeployed to the South Western Division.

The Southern Region examined the case for the replacement of steam on the Bournemouth/ Weymouth routes during 1962. Although the financial case had narrowly favoured dieselisation by use of both DEMUs and locomotives, the management of the time nevertheless decided in favour of electrification. It was clear that the original intention to electrify all the branches leading from the Bournemouth main line could no longer be justified, and that the electrification scheme as a whole would be pruned. At the same time, an exercise was in progress to see what could be done about the Region's uneconomic passenger services west of Exeter; in fact, these SR lines were shortly to be transferred to the WR which inherited the problem.

As at the autumn of 1962, the Bournemouth electrification project was envisaged as including the main line via Southampton to Bournemouth (from Sturt Lane Junction, Brookwood), the Alton-Winchester line and the Lymington branch. The line from Bournemouth to Weymouth, the Basingstoke to Salisbury stopping service, and the Swanage branch would be dieselised using DEMUs. Waterloo-Southampton and Waterloo-Weymouth Quay boat trains would be hauled by electro-diesel locomotives but all other weekday passenger trains on these lines would be EMU-worked. The Waterloo-Salisbury-West of England expresses would be cut short at Exeter but would remain steam-worked although diesels normally employed on freight services would be used to work summer weekend holiday trains. The Somerset & Dorset line was excluded from forward planing because it was, to quote SR official reports, 'under evaluation for closure'. If work commenced on the Bournemouth electrification project early in 1963, it was claimed that completion could be expected by mid-1967.

The SR's plans were soon overtaken by a late 1962 BRB directive that the future of many lines and stations must be examined as a matter of urgency. Also, the order was that rolling stock fleets must be thinned out, with the result that a large number of SR steam locomotives were withdrawn at the very end of 1962, including all the remaining ex-LBSCR 'K' 2-6-0s

Below:
Oxted, February 1960. A Tunbridge Wells West-Victoria train approaches behind BR '4' 2-6-4T No 80149 while a train to Tonbridge via Hever departs behind 'L' 4-4-0 No 31762, instead of the usual 'H' class 0-4-4T working push-pull. *Derek Cross*

Above:
The sight of engines dead at locomotive sheds and awaiting removal to scrapyards was commonplace in the 1960s. This is Redhill shed (on the left of the picture) on 10 March 1963 when it had withdrawn 'N1' 2-6-0s and 'Schools' No 30930. Setting off for Tonbridge with the 3.11pm from Redhill is 'N' 2-6-0 No 31873.
G. D. King

and 'Schools' 4-4-0s. The SR coaching stock allocation was also trimmed considerably, including no less than 38 catering vehicles, some of which were far from life-expired. Coincidentally, the interior condition of much of the SR's rolling stock deteriorated from this time.

Eventually, a revised version of the Bournemouth electrification was approved by Government in September 1964 but the Alton-Winchester line was excluded. Like the earlier scheme it was due for completion in 1967. Other changes included the use of diesel locomotives displaced from elsewhere on the SR with the arrival of a new batch of electro-diesels. Completion of the electrification was planned to allow the SR to withdraw its last 250 steam locomotives.

The clearance of the remaining steam locomotives was made easier by the transfer to the WR of the Salisbury-Exeter line, from west of Wilton, and of the decision made in the spring of 1964 that the services on the main lines to the West would be rationalised. The former SR through trains to Exeter and beyond would now terminate at Exeter St Davids and become semi-fast services worked by diesel-hydraulic locomotives based on the WR. Apart from Exeter-Exmouth, the branch lines of East Devon were closed to all traffic.

Of the remaining SR steam-worked lines, the Reading-Redhill passenger trains had been considered for closure under the Beeching report but, in the end, a rationalised service was introduced in January 1965. This used makeshift trains formed of a spare Hastings DEMU power car and trailer and EMU driving trailer - what became known as 'Tadpole' sets.

Typical of the steam-worked SR branches existing in the early 1960s was that from Three Bridges to East Grinstead. Its main function was as a feeder service between the Brighton main line and the Oxted lines. Improvements to the Oxted line services had come in the summer 1955 timetable and the introduction of an hourly interval off-peak service between Victoria, East Grinstead and Tunbridge Wells West. The weekday trains connected at Oxted into/out of a motor-train which served stations to Tonbridge via Edenbridge, and another at East Grinstead for Three Bridges, from where there were connections to Brighton and Bognor. It was all part of the SR's commonsense approach to passenger services, derived from its experience with the prewar electrification by which the regular interval timetables had generated increased patronage, and interlinking connections were a major feature.

The Three Bridges-East Grinstead service was worked by what the SR called 'motor-trains', push-pull workings with a tank engine and a couple of pre-Grouping coaches, the brake vehicle being adapted to provide a cab with controls for the driver when the engine was propelling the train; the fireman remained on the engine to fire. The 1955 timetable changes had called for an increase in motor-train working, with the result that some ex-LSWR 'M7' 0-4-4Ts were reallocated to Three Bridges shed to cover the East Grinstead workings.

Ex-SE&CR 'H' 0-4-4Ts had worked over the East Grinstead-Three Bridges line from the late 1920s but it was not until 1956 that they were allocated to Three Bridges to share duties with the 'M7s' on the East Grinstead trains. The last 'H' to work a motor-train on this line was No 31263, now preserved on the Bluebell Railway, and it made its final workings in January 1964. Afterwards DEMUs took over but the line was later closed completely. The pre-Grouping

motor-train sets had been withdrawn from 1960, replaced on the Three Bridges-East Grinstead and other workings by SR Maunsell corridor stock which had converted to form two-car sets. On some lines, such as Horsham-Brighton and Horsham-Guildford, motor-train working ceased in 1961 in favour of ordinary loco-hauled coaches hauled by Ivatt '2' 2-6-2Ts transferred from the LMR.

Some of the Maunsell locomotive classes lasted almost until the end of steam on the SR. 'N' class Moguls in particular found new homes further west with the completion of the two stages of Kent Coast electrification and the arrival of diesel-electric locomotives. Many of the 'N' and 'U' class engines had been extensively renewed from 1955 with new front ends, or even complete new main frames, new cylinders and chimneys and blastpipes. Some of both classes survived into 1966, and appeared on at least one rail tour.

Another durable Maunsell type was the 'S15' mixed traffic 4-6-0, the last of which dated from 1936. The first withdrawals had occurred in 1962 but many were hard at work into 1965. No 30837 was kept serviceable until January 1966 when it worked a well-remembered rail tour from Waterloo to Eastleigh via the Mid-Hants line.

With the transfer of BR Standard classes and Ivatt 2-6-2Ts to the Southern Region, much of the character of train working in the Southampton/Bournemouth/ Salisbury areas was changed but the SR enginemen, in contrast to some footplate crews elsewhere, obtained some excellent work from the Standard engines, although some of those engines transferred from other Regions in later years had been in poor shape. Late in 1963, the BR 2-6-4Ts

Studies in engine front ends

Above left:
Now preserved 'Merchant Navy' Pacific No 35027 *Port Line* arrives at Wareham with the down 'Royal Wessex' Waterloo-Weymouth express on 4 August 1966. In the background stands BR '4' 2-6-4T No 80138 with the branch passenger train for Swanage. *Verdun Wake*

Left:
Two Bulleid 'Q1' 0-6-0s, Nos 33015 (nearest) and 33036 at Guildford shed on 23 May 1964.
J. R. Besley

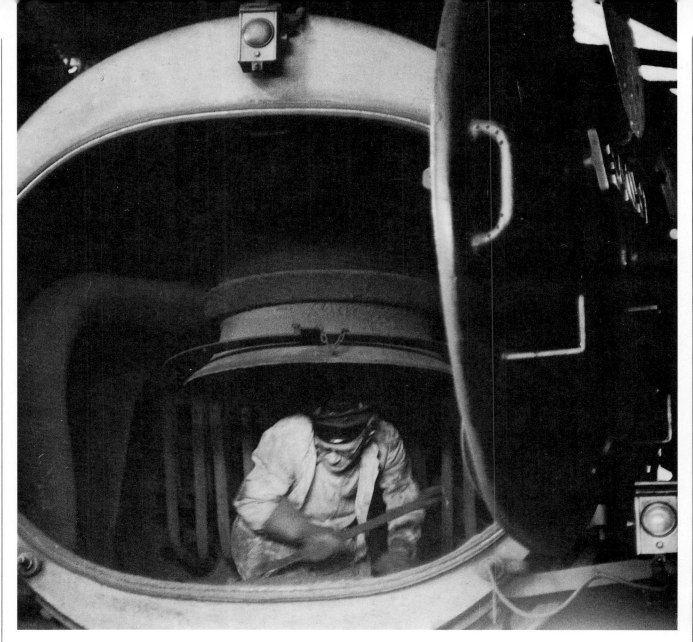

were put to work on the S&D line working stopping trains over the 71 miles from Bournemouth to Bath. In the London area, these engines took over from 'M7s' on local freight and empty stock workings although the Maunsell 'W' 2-6-4Ts could be seen from time to time until they were withdrawn in mid-1964. With the closure of the Hayling Island branch late in 1963, the last 'Terrier' 0-6-0Ts were withdrawn, and this marked the end of ex-LBSCR engines in SR service.

Modern steam power was being run down. The unrebuilt Light Pacifics had been withdrawn from June 1963, the first two 'Merchant Navy' Pacifics were withdrawn in February 1964, followed shortly afterwards by the first rebuilt Light Pacifics. The last Bulleid Pacifics to be given heavy repairs had been overhauled at Eastleigh Works early in 1964.

Otherwise, apart from diesel and electric traction, Eastleigh was concentrating on the overhaul of BR Standard classes. The works dealt with those on the SR's allocation and some from other Regions. The latter were either run in on SR services or otherwise 'borrowed', as were the LMS '5s' and others that arrived on the SR with through trains such as the York-Bournemouth express. Early in 1964, thoughts had turned to transferring some of the 'Duchess' Pacifics from the LMR but this was rejected because of the limited clearances at some locations.

During the summertime, steam operations on the South Western Division were reduced with the use of Type 3 diesels on express workings as train heating was not required then; these locomotives (later Class 33) had electric train heating only, and most SR Standard stock and all Bulleid coaches were steam-heated only.

The final run of the 'Atlantic Coast Express' had taken place on 4 September 1964 when, complete with train headboard, 'Merchant Navy' No 35022 *Holland America Line* had turned in an excellent performance with a 13-coach train. Although the express workings to Exeter were turned over to diesel-haulage, steam remained in use on some stopping trains between Waterloo, Basingstoke and Salisbury. With the ending of steam working on the Waterloo-West of England expresses in September 1964, more 'Merchant Navy' Pacifics were available for the Bournemouth line; in practice, many of the turns were handled by rebuilt and unrebuilt Light Pacifics and BR '5' 4-6-0s.

The SR received an allocation of 2-6-2Ts from the LMR in late 1965, with a view to their use on the Isle of Wight as stopgap power once they had been suitably modified to suit the restricted loading gauge of the Island's railways. However, it was decided to soldier on with the ex-LSWR 0-4-4Ts and to electrify the one line, from Ryde to Shanklin, that was being retained. On the mainland, the Swanage and Lymington lines were the last branches on the SR to have steam-hauled passenger trains.

One unusual working on the otherwise steamless Central Division was of the Plymouth-Brighton through train which for part of 1966 was steam-worked from Salisbury to Brighton and back, usually by a Bulleid Light Pacific. From the beginning of that year, a Brush Type 4 diesel (later Class 47) was used to work the 'Pines Express' through to Poole, and from September 1966 the SR received some of this class on loan for its principal express duties. From January 1967, the 'Bournemouth Belle' was booked to be diesel-hauled but, in practice, steam appeared on a number of occasions. Indeed, from 2-9 July 1967, in the last week of the train's existence, on at least three occasions the 'Belle' was steam-hauled in one direction.

By January 1967, the SR's allocation of steam power was down to just 77 engines, mostly Bulleid Pacifics but 10 or so of each of the BR '5' 4-6-0s and '4' 2-6-0s, and not forgetting the lone BR '3' 2-6-0 No 77014 which during the previous year had been transferred from the LMR.

With the approaching end of steam traction, many remarkable runs were recorded with steam-hauled trains on the Bournemouth and Salisbury lines, particularly with the Bulleid Pacifics. Some of the claimed speeds were over 100mph - way above the official speed limit. Having said this, after 1965 much of the day-to-day performance on the Bournemouth

Below:
The down 'Bournemouth Belle' behind steam power on 10 September 1966. Passing Farnborough is 'Merchant Navy' No 35012 *United States Line. D. A. Idle*

line had been anything but good, for many engines were in poor condition, increasingly grubby and the Pacifics devoid of their nameplates. The rolling stock was often unkempt and in similarly poor condition.

None the less, enthusiasts and many other well-wishers were keen to mark the passing of steam traction in the South. Quite apart from enthusiast rail tours, the SR marked the end of steam by running officially sponsored special trains on 2 July, one to Weymouth (with No 35028 *Clan Line*), the other to Bournemouth (with No 35008 *Orient Line*). The return working of the second train was double-headed, with No 35007 *Aberdeen Commonwealth* used from Weymouth to Bournemouth.

Sunday, 9 July 1967 was the last day of steam traction on the SR. The final express working was the 14.07 Weymouth-Waterloo hauled by 'Merchant Navy' No 35030, the former *Elder Dempster Lines*, which had taken the place of a diesel. Steam was used from Weymouth to Westbury for three inter-Regional freight trains. The last local turns took place in the Bournemouth area, and the very last steam-hauled revenue-earning train of all on the SR was a Bournemouth-Weymouth van train worked by 2-6-0 No 77014.

The Eastern and North Eastern Region in the 1960s

East Anglia had been one of the areas chosen by the BTC to change completely from steam to diesel and electric traction. By the summer of 1961, the 'Britannias' were making their last runs on Liverpool Street-Norwich and Liverpool Street-Clacton expresses and soon gravitated to freight working from March depot before being transferred to the LMR. The last scheduled steam train to appear at Liverpool Street had been a boat train working from Harwich

Parkeston Quay, headed by a 'B1' 4-6-0 in September 1962. Steam clung on to a few freight duties from March shed before ceasing entirely at the close of 1963. The London, Tilbury & Southend section's passenger service had been completely dieselised or electrified in the summer of 1962.

Elsewhere on the ER, in the first couple of years of the 1960s plenty of steam remained at work, particularly in South Yorkshire, although the dieselisation of freight turns had begun in earnest during 1962. Initial dieselisation had been in the Sheffield area but, with the opening of further diesel depots, it had extended to trunk freight services on the GN main line from April 1963.

Although numerous Type 4 and the 22 'Deltic' Type 5 diesels had been delivered to cover express passenger trains on the East Coast main line, Pacifics allocated to ER and NER depots were diagrammed to work King's Cross-Leeds trains and some additional Newcastle turns into the first half of 1963. The arrival of 30 Brush Type 4s during late 1962 and early 1963 was sufficient to allow steam express power to be eliminated although for the next year or so some 'A1s', 'A3s' and 'A4s' remained in use on the ER, usually as substitutes for diesels or on menial work. Some of the ER 'A4s' were transferred to the Scottish Region for further express passenger duties.

In April 1963, 'A4' No 60022, the record-breaking *Mallard*, had worked its final express train for BR. 'A3' No 60103 *Flying Scotsman* had headed a King's Cross to Leeds express for the last time on 14 January 1963, having already been purchased for preservation

Below:
Diesels were beginning to rob LNER Pacifics of many regular East Coast express passenger turns in 1960. 'A4' No 60032 *Gannet* crosses the viaduct at Relly Mill, near Durham with the Saturday 9am King's Cross-Edinburgh Waverley on 23 July 1960. *Ian S. Carr*

Above:
An outpost of steam shunting engines. Ex-Midland Railway '1F' 0-6-0T No 41708 (now preserved) raises the echoes at the head of a train of pig-iron at Staveley Iron Works, near Chesterfield, in December 1963. *Colin Boocock*

by Alan Pegler. The end was not far away for express steam power and June 1963 brought the elimination of all steam working in the King's Cross area, and closure of the famed King's Cross steam shed.

For a little longer, steam-hauled trains travelled as far south as Hitchin on one or two freight workings, but the end of steam on the GN main line was not long delayed. Working progressively northwards, ER steam sheds were closed down and, by June 1966, the Region no longer possessed a steam allocation. Latterly, the engines used had been 'B1' 4-6-0s, Austerity 2-8-0s, 'O1' 2-8-0s and '9F' 2-10-0s.

The East Coast main line north of Doncaster featured a number of steam workings into 1965. The last Pacific assigned to a main line passenger working was 'A1' No 60145 *Saint Mungo* which worked from York to Newcastle and back with a relief train on 31 December 1965. This was over what was still a North Eastern Region line but, during 1965, proposals to amalgamate the Eastern and North Eastern Regions was well in progress, and took effect from New Year's Day, 1967.

While the former Eastern Region had dispensed with steam, in the Northeast and parts of West Yorkshire it continued to earn its living on freight services. Passenger duties were few and far between, and then usually only as a result of diesel failures.

The North Eastern Region had made extensive use of DMUs to displace semi-fast and local steam-worked passenger services. The Region's schemes had included the trans-Pennine Hull-Liverpool services, dieselised in 1961, and that encompassing the Calder Valley main line and continuing to Manchester which went over to DMU operation in March 1962.

After that date, there were few regularly steam-worked local trains on the NER, other than those coming to York from Sheffield, and to Leeds from west of Skipton. Another steam outpost in the West Riding involved the use of NER-based 2-6-4Ts, mainly on the Bradford and Harrogate portions of King's Cross-West Riding expresses. These and seasonal trains continued to be steam-hauled until 1967.

It was a different matter with the Northeast's freight trains, well into 1966. On coal and mineral trains serving Tyneside, Teesside and Wearside there were numerous ex-NER 0-6-0s and 0-8-0s at work, supplemented by postwar 'K1' 2-6-0s and Ivatt 2-6-0s, as well as '9F' 2-10-0s, the last-mentioned (just) clinging to their duties on the Tyne Dock-Consett

Above:
On 2 June 1966, 'B1' 4-6-0
No 61388 pulls out of Bradford
Exchange with the 11am through
coaches to King's Cross which will
be combined at Wakefield with the
main part of the train from Leeds
City. In the background is the last of
the Fowler 2-6-4Ts, No 42410.
Malcolm Dunnett

Left:
'9Fs' at front and back as a Tyne
Dock-Consett iron ore train climbs
between Pelton and Beamish on
29 April 1964. No 92063 is at the
front, with No 92098 banking on
this steep stretch of line.
Malcolm Dunnett

iron ore trains. On West Riding lines, such as the Calder Valley main line, as late as January 1967 steam traction from LMR sheds and the remaining locally based locomotives still accounted for 50% or so of freight working.

The first half of 1967 saw steam sheds at York, Hull, Wakefield and South Blyth lose their steam allocations and, by the end of July 1967, no more than 124 steam engines were in stock on the enlarged ER; four of them were 'Q6' 0-8-0s and seven were 'J27' 0-6-0s. With the closure of steam depots at Sunderland and West Hartlepool that September, steam became extinct in northeast England and BR's

last pre-Grouping engines were taken out of service. For some time longer, industrial - mainly National Coal Board - engines could be seen manfully keeping the tradition of steam alive, at least until February 1969 and the end of 'main line' steam operations from NCB Philadelphia, Co Durham.

October 1967 brought the closure of Holbeck (Leeds), Low Moor (Bradford) and Normanton depots as well as the end of several classes - the Stanier and Fairburn 2-6-4Ts, the 'Jubilee' and 'B1' 4-6-0s, Austerity 2-8-0s and 'K1' 2-6-0s. On 4 November 1967, '8F' 2-8-0 No 48276 from the ER's final steam shed of Royston, near Leeds, powered the 15.00

freight from Carlton North Sidings, near Barnsley, to Goole and returned light engine. This was the final scheduled duty of an ER-allocated steam locomotive. LMR engines continued however to work into the area from across the Pennines.

After restoration for preservation, Alan Pegler's 'A3' Pacific, now running as LNER No 4472 *Flying Scotsman* was passed to work over BR main lines. It could be seen at the head of a number of specials in the mid and late 1960s. After November 1967, and until the rule was relaxed with the approaching end of all BR steam, it was the only steam locomotive permitted to work specials on BR. In May 1968, No 4472 ran nonstop with an excursion from King's Cross to Edinburgh.

The official attitude to steam-hauled specials had been a very different matter on the Eastern and North Eastern Regions in the early/mid-1960s. As late as May 1964 *Flying Scotsman* and another 'A3' had worked an enthusiasts' excursion from King's Cross to Darlington and back, despite the otherwise total ban on steam south of Peterborough. There were Ian Allan Locospotters' special trains to Doncaster Works of the early 1960s which made use of all sorts of famous steam motive power such as *City of Truro*, Midland Compound No 1000 and then what was just another engine in BR's stock - No 60103 *Flying Scotsman*.

This is a reminder that like all the remaining 'A3s', by then *Flying Scotsman* had been modified with a Kylchap double blastpipe and double chimney. Most of the class, including *Flying Scotsman*, had been further changed in appearance with the fitting of German-pattern trough, or 'elephant ears' smoke deflectors. During the late 1950s and early 1960s, those LNER-design Pacifics allocated to King's Cross depot were kept beautifully clean and were in excellent mechanical condition. They were putting up some of their best work since the prewar LNER heyday of the 1930s.

The Scottish Region in the 1960s

From occupying a place near the bottom of the Regional league table for diesel traction, the ScR soon moved near the top. This resulted from the BTC's acceptance in 1957 of a ScR commitment to breaking even by 1962 - an optimistic target if ever there was one! By that time, the ScR expected to be using diesel traction on all its branch lines, and almost all its main lines. But dieselisation was seen as a stopgap only for there were plans for extensive electrification, including the Perth-Inverness line. The first major dieselisation scheme to meet the 'break even by 1962' commitment was the £5 million project approved in August 1957 that aimed to eliminate steam from former Highland and Great North of Scotland lines.

Below:
Steam working on the West Highland line had less than two years to run when 'K1' 2-6-0 No 62012 caught the eye of the cameraman as it passed Banavie on 6 April 1960 with a Mallaig-Fort William train. *D. M. C. Hepburne-Scott/Rail Archive Stephenson*

The Region was however near the top of the list of suburban electrification schemes and it was forecast that both the north of the Clyde and the south of the Clyde networks would be electrified by 1962. This 1956 forecast was fulfilled exactly, despite early problems with the new electric trains in 1960 that had enforced a short-term reversion to steam working.

Also in 1957, the ScR together with the Eastern and North Eastern Regions agreed to a £3 million scheme for the acquisition of 23 'Deltic' locomotives (later amended to 22) to take over the most important East Coast passenger workings from Pacifics. The ScR allocation of 'Deltics' was eight locomotives which were to be based at Haymarket depot. The expected delivery date was 1959 but this was not achieved.

One feature of the ScR from the late 1950s was the way in which surplus steam locomotives were placed in open storage at various locations. Among the classes first treated in this way were the 'D11' and 'D34' 4-4-0s, then more modern steam locomotives including 'Jubilee' 4-6-0s, 'A3' Pacifics and 'V2' 2-6-2s began to be stored. At one stage, 70 engines languished at Bathgate. From late 1959 the ex-Caledonian Pickersgill 4-4-0s suffered from heavy withdrawals but, even into 1961, LMS '2P' 4-4-0s were at work on local passenger trains in Ayrshire.

Some lightly used ScR branch services, such as Crieff-Comrie, were dieselised from 1958 using four-wheeled diesel railbuses. The Region also received a number of diesel shunting engines to replace steam in docks and other industrial locations. Further steam strongholds to be dieselised included the Glasgow-Ayr-Girvan-Stranraer line where DMUs took over in

late 1959 and the Aberdeen-Inverness fast trains, converted to DMU working in July 1960.

The DMU depot at Leith was the base for the Region's first main line diesels when these began to arrive in late 1958. The first examples had been delivered originally to the ER but from 1959 Type 2s were sent directly to Scotland, both for the Highland and north of Aberdeen schemes. From August 1959 the Type 2s were used in multiple on Edinburgh-Aberdeen express trains, displacing Haymarket Pacifics to other work. Then, in 1960, the new diesel depot at Haymarket began to receive Type 4 diesels for work on the East Coast main line and Edinburgh-Aberdeen; again, it was ex-LNER Pacifics and 'V2s' that were replaced. In time, all principal West Coast express duties were covered by LMR-based diesels.

By the summer of 1961, nearly all workings on the former GNSR lines were diesel-worked (except for the Tillynaught-Banff branch) as were the ex-Highland lines north of Inverness. In the western Highlands, ex-CR 0-4-4Ts were still at work on the Ballachulish and Killin branches in the summer of 1961. Next it was the turn of West Highland line steam to give way to diesels which had taken over most workings by late 1961. By the summer of 1960 diesel locomotives were working through to Wick with trains from Inverness.

From the late 1950s, ex-LNER Pacifics and 'V2' 2-6-2s had increasingly appeared on the former Caledonian Railway main line via Forfar to Perth at the head of express passenger trains and the up West Coast Travelling Post Office special from Aberdeen. From 1960, ex-LNER Pacifics were also working into

Right:
One of the engines returned to steam by the Scottish Region was Highland Railway Jones Goods 4-6-0 No 103, here being used on an enthusiast tour train of June 1960 and seen working between Aviemore and Kincraig on the Highland main line.
W. S. Sellar

Glasgow St Enoch with Midland line expresses from St Pancras and Leeds that were routed over the Settle & Carlisle line and then via Dumfries and Kilmarnock.

The North British Type 2s had begun operations on the Glasgow-Perth main line but soon proved problematical. Although at work in some numbers during mid-1961, it was not long before once more they were mostly laid aside, and early in 1962 steam power was in charge of the Glasgow-Aberdeen trains, including LNER Pacifics from Haymarket and Ferryhill, Aberdeen depots. In February 1962, 'A4' No 60031 *Golden Plover* was transferred from Haymarket shed to the former Caledonian/LMS shed at St Rollox, Glasgow, and, in the same month, Haymarket's No 60027 *Merlin* had completed a Glasgow Buchanan Street-Aberdeen test run to a 3hr schedule.

Rumours soon spread that some Glasgow-Aberdeen expresses would shortly be accelerated to an overall 3hr schedule and would be steam-worked. This speculation was soon confirmed by the ScR, and the initiative to use LNER Pacifics on these trains is believed to have come from the Regional General Manager of the time, James Ness. The 3hr Glasgow-Aberdeen trains, featuring tightly timed schedules, were to be the swansong of the 'A4s', and ran over a line scheduled for closure to passenger traffic under the Beeching Plan. This was the attractive Strathmore route from Stanley Junction, north of Perth, through Forfar and to Kinnaber Junction.

Running Mondays-Saturdays only, there were just two 3hr trains each way between Glasgow and Aberdeen and they began running from September 1962. By the following year, the days of the 'A4s' in regular service south of the Border were drawing to an end, in particular on the Eastern Region following the closure of King's Cross shed during June 1963.

Some of this shed's 'A4s' were transferred first to New England and then, with some of their compatriots from Gateshead, sent to Scotland during the autumn of 1963, some initially going into store. For the next three years Pacifics could be found at work on the 3hr Glasgow-Aberdeen expresses although diesels took over one pair of trains in 1964, and from time to time Class '5s' substituted for the 'A4s'.

Rail tours elsewhere on BR called upon the services of the Scottish 'A4s', such as No 60024 *Kingfisher* which worked from Doncaster to Edinburgh and back in May 1966, this rail tour raising funds for the preservation of 'A4' No 60007 *Sir Nigel Gresley*. Early in September 1966, the 'A4s' made their last appearances on the Glasgow-Aberdeen expresses. The ScR marked the event by running a special on 3 September 1966 behind No 60019 *Bittern*, and a £2 fare bought enthusiasts a 3hr run in both directions between Glasgow and Aberdeen.

Early in 1962, the Scottish Region provoked a public storm and strike threats from railway trades unions when it was announced that line closures were being considered, principally in rural areas but also including outer-suburban lines, and there would be a general thinning out of train services. Dr Beeching's *Reshaping of British Railways* report seemed to include a disproportionate amount of Scottish railways in its lists for closure. In the event, although a number of lines succumbed, much of the system was reprieved.

Unhappily, the so-called 'Port Road', from Dumfries to Challoch Junction, near Stranraer, was one line that closed completely. It was a remarkable single line

railway that passed through beautiful scenery, perhaps best seen on a spring morning when dawn was breaking as the overnight 'Northern Irishman' sleeping car express forged cross-country to Stranraer.

First the branch lines connecting with the Port Road were closed then, on 14 June 1965, services were totally withdrawn between Maxwelltown, the first station out of Dumfries, and Challoch Junction, including the Castle Douglas-Kirkcudbright branch. These were steam-worked lines to the end, with the two- and three-coach Dumfries-Stranraer stopping trains worked by LMS 'Black Fives' or BR '4' 2-6-4Ts. With closure of the Port Road, the 'Northern Irishman' was rerouted via Mauchline, Ayr and the other route to Stranraer, via Girvan. A trip on the sleeping car express, double-headed by a 'Britannia' and a 'Black Five' over the fearsome grades of its new route, was never to be forgotten.

Both lines to Stranraer had featured regular workings of the BR Standard 'Clan' class Pacifics which numbered no more than 10. By 1962, the first was condemned but the last of the class survived at Carlisle Kingmoor shed until 1966.

By the mid-1960s Scottish steam engines old and new were being rapidly withdrawn. Until the very end engines of one pre-Grouping Scottish railway were at work hauling coal and mineral trains in the Lowland belt and in Fife. The last ex-Caledonian 0-6-0s had been withdrawn at the end of 1963 but North British 0-6-0s continued until the end came on May Day 1967 when the ScR withdrew its steam allocation.

The London Midland Region in the 1960s

Always the largest of the BR Regions, the LMR stretched from South Acton to Carlisle, from Holyhead to Melton Mowbray. Yet, at 4,177 route-miles in 1958, it just took second place in extent to the Western Region. The LMR had six Divisions - London, Birmingham, Nottingham, Manchester, Liverpool and Barrow - important shipping services, some of BR's largest stations, some of its longest tunnels and its major bridges and viaducts. Eight of the BR workshops came under its control, and it operated important shipping services across the Irish Sea.

By the late 1950s, all the Region's energies were being concentrated on the first stages of the 25kV electrification of the main lines south from both Manchester and Liverpool, and other modernisation projects on the former North Western main line.

The LMR had taken a look at itself in 1959 and decided it must become both smaller and more tightly managed. Nearly one-third of the nearly 1,200 stations were to be closed, and the Region concluded that a large number of branch lines were suspect. So, too,

were some sections of main line, particularly the former Great Central and the Settle & Carlisle. A few lines listed for closure under the Beeching Plan had been intended for dieselisation, but in the event stayed steam-worked until they were closed to passengers, examples being Northampton-Kettering-Melton Mowbray-Nottingham and Nottingham-Worksop.

The scale of the closures and changes to services becomes apparent on realising that, after the publication of the Beeching report, the LMR alone shut 371 passenger stations, withdrew no fewer than 55 passenger services and modified 35 others. As at 1968, another 24 services remained in operation awaiting Ministerial assent to closure. Route-mileage had been slashed to only 3,600.

Despite these closures and changes to services, the LMR of 1968 was by far the most unprofitable of the BR Regions. Admittedly, as from the beginning of 1963 it had taken under its wing former WR lines in the West Midlands and North Wales.

Below:
Demolition in progress of Euston's Great Hall during the summer of 1963. *Science & Society Picture Library NRM/GEN 0067*

Above:
The recently completed Carlisle Kingmoor marshalling yard seen in 1962, never to be used to its full capacity. *Ian Allan Library*

Through the early 1960s, both at the British Railways Board and at the LMR headquarters, planners were at work on the final stages for the rundown of the GC main line. The line had featured in Beeching's Reshaping Plan, but it was not until October 1964 that BR announced that it was proceeding with a five-stage scheme that would bring complete closure of 62 miles of the GC's 140-mile route south of Sheffield.

The main passenger workings on the GC were the Marylebone-Nottingham semi-fast trains, three of them each way, and in the summer 1963 timetable they were slowed by the addition of extra stops, the midday down train was retimed, and the departures of up trains from Nottingham Victoria tidied up.

For the first three years of the service Annesley and Neasden sheds had mostly used LMS 'Black Fives', with the appearance on occasion of 'B1' 4-6-0s borrowed from other depots, but by 1963 Annesley had an allocation of nine 'Royal Scots'. The depot continued to borrow locomotives from other sheds, including York-based 'V2s'. When one by one the 'Scots' were withdrawn, no other Class '7s' were found to replace them and so the Marylebone-Nottingham semi-fasts gradually reverted to Class '5' haulage.

One unstated reason for closure of the GCR main line was that it was intended to channel as much traffic as possible 'under the wires' of the West Coast main line on completion of its electrification. The plan had been that by the end of 1965 inter-Regional through freight services and the York-Bournemouth

express would have been diverted to other routes. The Calvert North Junction-Rugby Central and Culworth Junction-Banbury Junction sections of the GC route were to close completely, the surviving section north of Rugby being destined to carry a service to and from Nottingham Arkwright Street, operated by DMUs. Steam continued on the semi-fasts until the last day of operation on Saturday, 3 September 1966.

Elsewhere on the LMR, dieselisation of the Midland Lines got under way early in 1960 with the introduction of DMUs on the St Pancras-Bedford service. Soon afterwards, Type 2 and Type 4 main line diesels were appearing on the Midland Lines, and by 1962 practically all express trains south of Leeds and on the Bristol-Derby main line were in the hands of diesels. It took longer for steam to disappear from freight workings on these lines. North of Leeds, the express trains to Scotland using the Settle & Carlisle main line were dieselised by late 1961.

On the main line between Euston and Crewe, after 1959 the operation of express trains was at times disrupted by the slow work of electrifying this busy route. From November 1959, the popular Euston-Birmingham-Wolverhampton express service had been largely withdrawn, and to compensate the number of fast trains was increased on the WR

Above:
The shadows are closing in for 'Duchess' No 46233 *Duchess of Sutherland* as it nears Clifton & Lowther, on the climb from the north to Shap Summit with a mixed freight train on 13 April 1963. *John S. Whiteley*

Paddington-Birmingham route. The Euston-Manchester/Liverpool expresses were progressively dieselised from late 1959 although the apparent unreliability of the new motive power meant that for the next three years or so 'Duchess' and 'Princess Royal' Pacifics appeared from time to time.

Through steam working of freight and parcels trains and additional express passenger workings remained into 1964 when electric working had been extended as far south as Rugby. This once major railway centre was to lose nearly all its local services during the 1960s, with the closure of the lines to Leamington, Leicester Midland and Peterborough. Declining in importance, the steam shed remained, while to the south of the station lay the disused Rugby Locomotive Testing Plant as a reminder of times past.

A Stanier 'Duchess' Pacific at the head of a premier express was a fine spectacle associated with the heyday of the steam age West Coast route. One of the last regular turns worked by the class had been on the daytime Birmingham-Glasgow express but, by 1962, the Pacifics were mostly filling in on freight and parcels trains, acting as substitutes for diesels, and at holiday weekends showing their muscle on heavy additional express trains. No less than 13 of the class were withdrawn during 1963, and at one stage it looked as if all might be taken out of traffic with the close of that summer's timetable. By the spring of 1964, only 19 'Duchesses' were left, mostly stored to await summer traffic. Then it was proposed that some of the class might be transferred to the Southern Region to replace Bulleid Pacifics. The plan did not prove acceptable. Although the survivors put in some hard work during the 1964 summer timetable, all were withdrawn on its expiry. However, No 46256 *Sir William A. Stanier* was retained for a fortnight longer to work a rail tour. One reason for withdrawing the class on the LMR was that they were banned from working under the 25kV ac overhead electric catenary and to denote this fact acquired the dreaded yellow diagonal stripe on their cabsides. Fortunately, two 'Duchesses' and a 'Princess Royal' were purchased by Billy Butlin for preservation at his holiday camps.

The ex-LNWR 0-8-0s may have been less immediately appealing than the Stanier Pacifics, but equally they were time-honoured features of the railway scene. The 0-8-0s were also subjected to the yellow stripe. Four of the class were used by Bescot shed for a local freight turn until the end of 1964, and two of them were turned out to work a farewell rail tour on 12 December of that year.

Steam push-pull trains had tended to survive on lines that were threatened with closure and whose dieselisation was not considered worth while. By late 1964, the only surviving push-pull services were on the Western and London Midland Regions. The LMR's last but one steam push-pull was between Wolverton and Newport Pagnell, this being withdrawn on 7 September 1964. The very last was the service between Seaton, on the ex-LNWR Peterborough-Rugby line, and Stamford Town. The service finished its life as a steam push-pull working and closed on 2 October 1965 when the last turn was covered by Ivatt 2-6-2T No 41212 with two LMS-design non-corridor coaches.

Even in 1965 the enthusiast did not have to look far to see LMS-design coaches in service. Many were relatively modern and dated from 1949-51, and some were still in the process of receiving overhauls and being converted to dual steam/electric train heating. A number of LMS-design coaches came to be repainted in BR blue and grey livery, but by 1969 almost all had been withdrawn.

In the Northwest, apart from the growing use of DMUs, the steam age railway remained largely untouched until the late 1960s; indeed, a late 1967 trip north of Crewe to Preston was memorable for the transition from the electrically-worked southern part of the main line to steam-worked freight and parcels trains in the Warrington, Wigan and Preston areas. North of Preston, steam working continued on the West Coast main line over Shap to Carlisle until Carlisle Kingmoor depot closed to steam working at the end of 1967 which also marked the end of steam operations over the Settle & Carlisle line. Carnforth then became the northern limit of steam working and Kingmoor shed was a 'dump' for withdrawn steam locomotives lined up to await their final movement to scrapyards.

Another landmark was reached in December 1967 with the withdrawal of all but one of the 'Britannia' Pacifics - this was No 70013 *Oliver Cromwell* which that February had received the last full overhaul carried out on a steam locomotive in BR's ownership - at Crewe Locomotive Works.

Steam working at Birkenhead and the use of '9F' 2-10-0s on the famed Bidston-Shotwick Sidings iron ore trains came to an end in November 1967; the

Below:
Fated to carry the yellow diagonal stripe, ex-LNWR 0-8-0s Nos 49361 and 49430 are prepared for evening duties at Bescot depot on 29 September 1964. *Dr D. P. Williams*

engine used on the last ceremonial working was No 92203 which was purchased for preservation by artist and railway enthusiast, David Shepherd. Buxton and Northwich depots closed to steam working at the start of March 1968.

By now, the BR steam locomotive fleet had been reduced to no more than 350 locomotives of just seven classes: Ivatt 2-6-0s, Stanier '5' 4-6-0s, Stanier '8F' 2-8-0s, the sole surviving 'Britannia', BR '4' and '5' 4-6-0s, and '9F' 2-10-0s. In January 1968, the largest steam allocations were at Carnforth (38 engines), Edge Hill (37) and Patricroft, Manchester (34). These were the remaining standard gauge steam engines but the LMR now also operated the Vale of Rheidol 2ft gauge line with its three 2-6-2Ts.

In May 1968, the last of any BR timetable to feature freight operations only) brought the closure of four

more steam sheds. All that were left were Newton Heath, Patricroft and Bolton, Carnforth, Lostock Hall (near Preston) and Rose Grove (near Burnley). Regular steam passenger working had bowed out when the Heysham-Manchester Victoria 'Belfast Boast Express' changed to diesel traction after 5 May 1968. On occasion, the 'Black Fives' used on this train had provided some sprightly running. After 1 July, only Carnforth and Rose Grove sheds remained to operate steam. Steam working gradually wound down and by 4 August, it was almost over for good - on 5 August all normal steam activity came to an end. Only BR's own commemorative steam-hauled special remained to run.

Early in 1968, BR had relented and allowed a number of steam-hauled rail tours to operate. This activity reached its climax on the last but one day of scheduled BR steam working - 4 August. Six steam-hauled specials toured Lancashire that day

Then came the final special train of 11 August 1968 - what at the time was expected to be the last steam . working on BR for all time - organised and run by the LMR to commemorate the end of standard gauge steam operations. BR seemed to have decidedly mixed feelings about its initiative, but then it had spent the previous few years clearing its network of steam power.

When it was announced that seats on the special train would cost 15 guineas (say, £140 in 1990s money) many enthusiasts accused BR of unfair profiteering. Advertised by posters which were headed 'British Rail runs out of steam', for their 15 guineas seat-holders were treated to 314 miles of steam haulage, and an at-seat service of lunch, high-tea and other refreshments. The train left Liverpool Lime Street behind now-preserved LMS '5' No 45110 which at Manchester Victoria was replaced by No 70013 *Oliver Cromwell* for the run to Carlisle via Blackburn and over the Settle & Carlisle line. For the return journey, a pair of 'Black Fives' were used throughout - Nos 44781 and 44871.

Above:
It's getting very near the end... on 29 December 1967, just 48hr before the end of steam working at Carlisle, a down freight train ascends Shap behind Stanier '5' No 44672, with BR '4' 4-6-0 No 75037 assisting at the rear. Behind No 44672 is a condemned Ivatt 2-6-0 which is en route to a scrapyard. *P. Weightman*

Below:
The bufferstops for steam at Rose Grove shed, near Burnley. *Malcolm Dunnett*

How steam ended the decade

Although the demise of BR steam had been long forecast, when the end actually came in August 1968 many people involved with railways felt a sense of unreality. It was difficult to believe that steam traction was missing from our railways. It wasn't of course .... for steam continued in harness at a number of industrial locations, on engineering trains for London Transport, while Alan Pegler, owner of *Flying Scotsman*, had adroitly negotiated an agreement with British Rail that promised to keep the LNER Pacific employed on occasional main line excursions into the early 1970s. Although BR had finished with standard gauge steam working, it retained the Vale of Rheidol 1ft 11½in gauge tourist railway connecting Devil's Bridge with Aberystwyth.

So main line standard gauge steam may not have survived to the conclusion of the decade, but not all the fires had been dropped on steam as a new decade approached, and the 1960s drew to a close. We finish therefore on a slightly quizzical note, at the same time neatly concluding the decade: *Flying Scotsman* on BR main lines, and a glimpse of the Vale of Rheidol. Even with the end of the 1970s, and then the 1980s and 1990s, the story of steam was - and is - far from finished.

Above:
It's 1 June 1969 and, heading along the West Coast main line between Thrimby Grange and Clifton & Lowther, south of Penrith, is No 4472 *Flying Scotsman* which is working a northbound special organised by the Gainsborough Model Railway Club.

State of the art diesel traction for the times was Class 50 diesel-electric, No D434, at the head of the 11.45 Glasgow Central-Birmingham New St express.

Arrangements were in progress for *Flying Scotsman* to visit North America and tour with a trade exhibition train. On 31 August 1969, *Scotsman* worked its last main line excursion on BR, before being shipped abroad. It was not to return until 1973 but that takes us into another decade. *Peter W. Robinson*

Below:
BR's steam fleet was restricted to the three narrow gauge locomotives on the Vale of Rheidol and a number of steam cranes, all of which kept steam expertise going on BR, into the 1970s and beyond.

From 1968, the Vale of Rheidol engines and coaches were repainted into BR's corporate identity colours and carried the BR totem, as seen here on No 8 *Llywelyn* which has just arrived at Devil's Bridge with the 13.30 ex-Aberystwyth on 22 July 1969. *G. F. Gillham*